Key Concepts in Geography

Key Concepts in Geography

Rodolfo B. Valdenarro

Delve
PUBLISHING
www.delvepublishing.com

Key Concepts in Geography

Rodolfo B. Valdenarro

Delve Publishing
224 Shoreacres Road
Burlington, ON L7L 2H2
Canada
www.delvepublishing.com
Email: orders@arclereducation.com

ISBN: 978-1-77469-146-5 (Hardcover)

ABOUT THE AUTHOR

Rodolfo B. Valdenarro Jr., is a Professor and licensed education practitioner at Centro Escolar University, Manila. He served as Director of Student Affairs and Services at Laguna State Polytechnic University from 2016-2019. He finished Master of Arts in Teaching Social Science at University of Rizal System and presently taking Doctor of Philosophy in Southeast Asian Studies at Centro Escolar University. He teaches Professional Education subjects, Social Sciences (History, Sociology, Humanities, Anthropology, Economics, Geography, Political Science) and Methods of Research. He served as Contributor on the 2011-2016 Philippine Youth Development Plan (PYDP) and President of State Universities and Colleges- Association of Student Affairs Administrators in Region IV (CALABARZON and MIMAROPA), Philippines from 2018-2019. Mr. Valdenarro has presented researches on Social, Behavioral, Educational and Gender Studies in national and international conferences.

TABLE OF CONTENTS

LIST OF FIGURES

LIST OF ABBREVIATIONS

AT	Accumulated Temperature
CESB	Central Electricity Generating Board
CEGB	Central Energy Generating Board
FAO	Food and Agriculture Organization
FSU	Former Soviet Union
LEDC	Less Economic Developed Country
NOX	Nitrogen oxides
WHO	World Health Organization
WTO	World trade organization

PREFACE

Geography is the field of science that deals with studying the diverse environments, places, and spaces of Earth's surface along with their interactions. It typically answers the questions of why things are as they are and where they are. The modern academic discipline of geography is deep-rooted in ancient practice. It is basically related to the characteristics or features of different places, especially their natural environments and peoples, together with the relations between the two.

Geography can help people understand the movement of the planets, their systems, etc. Concepts that are highly relevant to today's world, such as natural resources, climate change, water availability, and more, are much easier to understand by those who know geography well. In addition, the study of geography also aids people in understanding relationships between different cultures. In due course, this leads to a more accepting as well as culturally aware society.

Geographers explore both the physical properties of Earth's surface and the human societies that are spread across it. Additionally, they also inspect how human culture interacts with the natural environment. Geography pursues to understand where things are found, why they are at that specific place, and how they develop or change over a certain period of time.

A separate identity was initially named and formulated around 2,000 years ago by the Greeks, whose geo and graphein were combined to mean "earth description." Or "earth writing." Nevertheless, what is now understood as geography was elaborated before then, in the Arab world and elsewhere.

Ptolemy, the author of one of the discipline's first books, 'Guide to Geography (2nd century CE),' defined geography as "*a representation in pictures of the whole known world together with the phenomena which are contained therein.*"

People who study Geography have an exclusive perspective. One that comes with the acquaintance of many cultures as well as spatial awareness, which is not replicated in other disciplines. This combination of knowledge can support geographers in coming up with unique and effective solutions that others may not be able to observe.

Learning geography will help people to better understand broadcasts, support in fighting climate change, be a part of a global community, learn history and understand cultures in a better way. At the end of the day, geography will help people to become better global citizens overall.

This book evidently outlines key concepts of Geography in a highly accessible way that all geographers should readily be able to explain. It comprises excellent content for accessible overviews of key concepts across the vast field of geography.

Further, this book takes a completely different approach to present concepts of Geography. Instead of presenting readers with a comprehensive and novice-level introduction to geography, outlining places and vocabulary terms, this text approaches Geography as geographers actually recognize the discipline, emphasizing more on connections and an in-depth understanding of the core themes or concepts. We hope that reading the key concepts in Geography help the readers to better understand the terms and concepts.

A

Abiotic

The nonliving chemical and physical aspects of the environment that have an impact on the survival of living species and the overall ecosystem are referred to as abiotic components or abiotic factors in biology and ecology. Abiotic factors and the phenomena connected with them serve as the foundation for all of life.

Abiotic variables are non-living physical and chemical components that exist in an environment but do not interact with live organisms. Abiotic resources are often derived from the lithosphere, the atmosphere, and the hydrosphere, among other places. Abiotic variables include things like water, air, soil, sunshine, and minerals, to name a few. A substance that is not connected with or generated from live beings. Light, temperature, wind patterns, and precipitation are examples of abiotic variables that influence the environment.

Ablation

Ablation zone or ablation area refers to the low-altitude area of a glacier or ice sheet with a net loss in ice mass due to melting, sublimation, evaporation, ice calving, or other processes such as blowing snow, avalanche, and any other ablation. Ablation is a key part of the glacier mass balance.

Absolute Drought

The duration of 15 or more days in a row with less than 1 mm of rainfall on each of those days is described as dry spell; an absolute drought is defined as a period of 15 or more days in a row with less than 0.2 mm of rainfall on each of those days; and a partial drought is described as a period of at least 29 days in a row with a total rainfall total taking an average less than 0.2 mm of rain each day.

A typical criterion for determining whether or not a drought exists is when living creatures are suffering as a result of a lack of rainfall. In any event, this is a biological explanation instead of a meteorological definition, and it is far from acceptable due to the intricate and changeable interaction that exists between the weather, plants, and wildlife. More significantly yet, in today's urban culture, ideas of drought are frequently associated with the ability of the local water authority to cope with periods of low rainfall; this, in turn, is mainly dependent on the amount of water storage capacity available in the area.

Figure 1. Absolute drought.

Source: Image by LiveScience.com.

Absolute Humidity

At a given temperature and volume of air, absolute humidity is defined as the mass of water vapor divided by the quantity of dry air in that volume of air. The more water that can be absorbed by the air, the warmer it is. Absolute humidity is a measure of the amount of water vapor or moisture in the air that exists independent of the surrounding temperature. In the unit of measurement, grammes of moisture per cubic meter of air (g/m^3) are used.

At 30°C (86°F), the maximum absolute humidity of warm air is roughly 30 grams of water vapor per cubic meter of air, or 30 g/m^3. The greatest absolute humidity of cold air at 0°C (32°F), the freezing point, is roughly 5 grammes of water vapor per cubic meter of air, or 5 grammes per cubic meter of air. This would be the difference between the relative humidity in the summer months and the relative humidity in the winter months, respectively. Warmer air contains more water than colder air, but colder air holds less water than warmer air.

The ratio of the mass of vapor (moisture) in the carrier gas stream to the total mass of the carrier gas stream. As an illustration, 0.02 pounds of water per pound of air is used. According to the psychometric charts, this value may be used to calculate the relative humidity. Also helpful for calculating cumulative amounts in a stream owing to elements such as combustion products when a gas fueled heater is being utilized, evaporation, and ambient quantities, among other things. This is required in order to calculate the quantity of condenser or venting required.

Abstraction

When water is extracted from aquifers, the level of groundwater decreases. Over-abstraction occurs when the amount of water extracted is larger than the amount of water that falls as rain during a certain period of time. Water is being removed from the Sonoran Desert in Arizona for irrigation and development, causing the desert to become over-abstracted.

Currently, one of the most significant points of reference for modern human geography is a criticism of abstract thought or abstraction. The scope of this critique, on the other hand, has been constrained by the inclination to contrast the abstract with the experienced.

According to the author of this study, abstraction may be seen as an essential component in understandings of lived worlds in the process of construction. The relationship between abstraction and two things of disciplinary concern: experience and materiality, must be revisited in order to do this. A single technology of abstraction, the diagram, is used to bring these issues together before a positive criticism of abstraction for geographical reasoning.

Accessibility

It is commonly accepted in geography that the accessibility of a place is defined as the more-or-less great ease with which it may be accessed from one or more other locations by one or many persons who are capable of travelling using all or part of the currently available modes of transport. The accessibility is the capacity to get from one location to another in relation to another. The term accessibility means the ease with which people can get to their destinations. Those in more accessible areas will be able to reach activities and destinations more quickly than people in unavailable places. In a given length of time the latter will not be able to access the same number of sites.

Equal access and opportunity are determined by accessibility. In the United Kingdom, for example, the public transport accessibility level (PTAL) is a technique of transport planning that assesses the accessibility level of geographical areas in terms of public transportation.

Accumulated Temperature

It is the integrated surplus or deficit of temperature around a given datum that is measured as Accumulated Temperature (AT). In crop growth models, it is a metric derived from temperature data that may be used to predict crop growth. Accumulated temperature data may therefore be used by crop agronomists to simulate regional variations, such as those found across the United Kingdom, and may become more essential for estimating the impact of climate change in the coming years.

Previous research has demonstrated that understanding the response of a crop to a certain agronomic treatment regime permits the use of AT maps to demark regions favorable for crop development. Consequently, for temperature-sensitive crops such as maize, zones of appropriateness might be defined on a national or regional scale, depending on the situation.

Acid Rain

It is a broad concept that refers to any type of precipitation containing acidic components, such as sulfuric or nitric acid, that falls to or accumulates on the ground from the sky in either a wet or dry state. This can include acidic precipitation such as rain, snow, fog, hail, or even dust.

In the atmosphere, acid rain occurs when Sulphur dioxide (SO_2) and nitrogen oxides (NOX) are produced and carried about by wind and air currents. Sulfuric and nitric acids are formed when SO_2 and NOX combine with water, oxygen, and other molecules, respectively. These are then mixed with water and other elements before being released to the atmosphere.

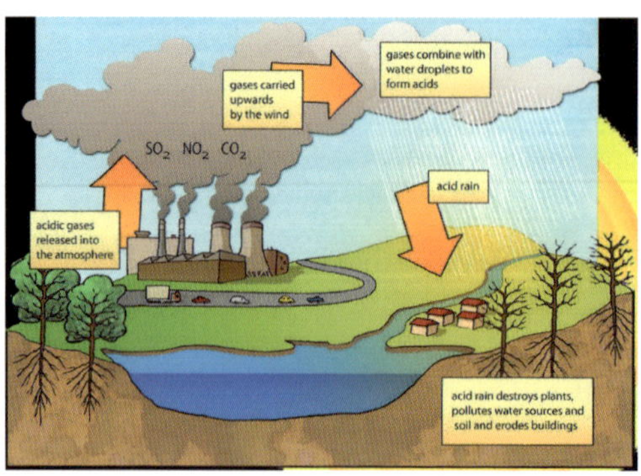

Figure 2. Acid rain.

Source: Image by flickr.com.

Even though some of the SO_2 and NOX that contribute to acid rain come from natural sources such as volcanoes, the vast majority of the pollution is caused by the combustion of fossil fuels. The following are the primary sources of SO_2 and NOX in the atmosphere-

- The combustion of fossil fuels for the generation of electricity. Electric power generators are responsible for two-thirds of the SO_2 and one-fourth of the NOX in the environment.
- The use of automobiles and heavy equipment
- Manufacturing, oil refineries, and other sectors are examples of this.

Acid Soil

Acidic soils are frequently referred to as sour soils because of their acidity. Acidity is defined as the characteristic of soils that have a pH level of less than 7 according to the scientific standard and a reading of 7 being neutral. The pH scale ranges from 0 to 14, with 7 corresponding to the midpoint of the scale. Alkalinity is indicated by readings greater than 7.

Gardeners may be shocked by the findings of a soil test, which may reveal that the soil on which they grow their crops has gotten more acidic since the last time they checked. Mother Nature, on the other hand, would be taken aback by your surprise. Acidic soil is common in many parts of nature, including most forested regions, and is the norm in many situations. That is why acidic soil is preferred by many forest plants, including the native North American ground cover bunchberry (*Cornus canadensis*). The Natural acidification of the earth occurs as organic matter decomposes whether in the yard or in the forest. The ground in or under the area where this decomposition is taking place naturally tends to become acidic. However, there are additional variables that might contribute to the soil being acidic, such as:

- The usage of chemical fertilizers is becoming increasingly common.
- Because of rainfall, minerals such as calcium and magnesium are lost to the environment.

Actinometer

Actinometers are tools that are used to measure the amount of heat emitted by radiation. Pyranometers, pyrheliometers, and net radiometers are all types of instruments that are used in meteorology to detect solar radiation. An actinometer is a chemical system or physical device that determines the number of photons in a beam either integrally or per unit time using a chemical reaction or physical device. When it comes to optical metrology or more broadly radiation metrology, actinometry is quite similar to radiometry; the distinction, however, lies in the fact that it does not relate to energies or powers, but to the number of radiation quanta or photons in case of optics. For example, the photon flux, which is the number of photons per second, is the actinometric quantity that corresponds to the radiometric quantity radiant flux in the radiometric space. The photon flux density i.e., photons per second and unit area) that corresponds to the radiometric quantity irradiance, which is also known as flux density, may be defined in the same way.

Active Layer

In both the anode and cathode of a PEMFC, a catalytically active layer is located near to the electrolyte membrane and serves as a catalyst. In addition, the layer is placed on a carbon paper that has been coated with PTFE, which serves as a current collector and gas diffusion layer, allowing for the passage of gas, water, and electrons. Platinum is the most active electrocatalyst when used in conjunction with pure hydrogen and oxygen. Nano-particles of platinum on a carbon substrate have been created in order to minimize cost, and the ongoing development of stable nano-structures continues to be an essential component of fuel cell research.

Nonetheless, one important element of electrocatalyst development is the creation of anode catalysts for operation on reformed hydrogen produced by the reforming of methanol, natural gas, or gasoline as a fuel. This is due to the fact that, while gas-diffusion electrodes with a loading of 0.1–0.2 mg cm^2 of dispersed platinum on carbon show very small polarization losses when operating on pure H2, these losses are increased to unacceptable levels when even small amounts of CO (10–100 ppm) are present.

In fuel mixes derived from reformed natural gas or alcohols, this is always the case, unless severe steps are taken to decrease the CO concentration to acceptable levels throughout the fuel processing process. Many attempts have been made to understand the process of CO poisoning, employing a wide range of experimental techniques to achieve success. It is commonly accepted that CO poisoning occurs as a result of a significant adsorption of CO on the catalyst surface, which prevents the hydrogen adsorption phase from taking place properly.

Adaptive Radiation

The development of a group of animals or plants into a diverse range of kinds that are well-suited to different forms of existence is known as adaptive radiation. It is among closely related groups that have developed in a very short period of time that adaptive radiations are most demonstrated. A prominent example is the evolution of primitive mammalian stock into species that are capable of running, leaping, climbing, swimming, and flying, which began during the Paleogene Period which began 66 million years ago and continued until the present. Other examples include marsupials from Australia, cichlid fish, and Darwin's finches, among others also known as Galapagos finches.

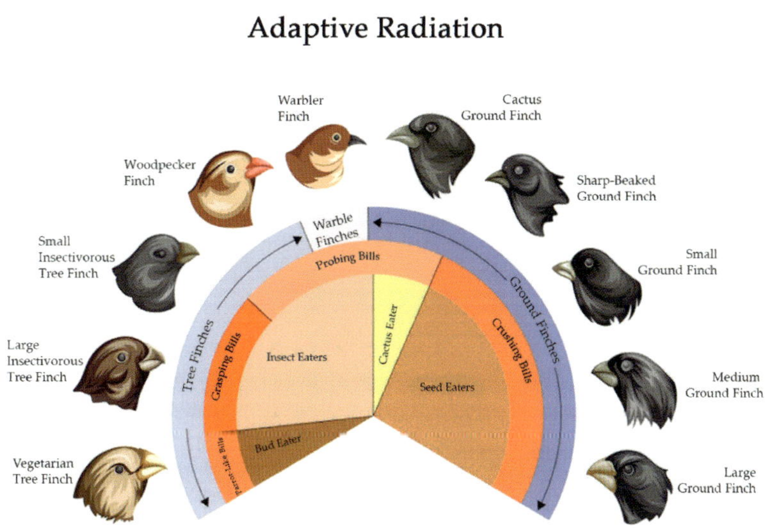

Figure 3. Adaptive radiation.

Source: Image by biologydictionary.net.

In archipelagoes that are geographically separated from the mainland, several examples of speciation through adaptive radiation may be discovered. In addition to the Galapagos Islands, the Hawaiian archipelago, with its several volcanic islands and relatively small total land area, is home to an astounding number of plant and animal species that are endemic, meaning that they have evolved only in Hawaii and cannot be found anywhere else on the planet. The Galapagos Islands are the most well-known example of this. More than 90% of the native species of Hawaiian floral plants, land mollusks, birds, and insects are endemic, meaning they can only be found on the island.

Adsorption

The process in which molecules of a material from a gas mixture or liquid solution get attached to the surface of a solid or liquid is known as adsorption. The material that is being absorbed is referred to as an adsorbate, while the substance that is absorbing is referred to as an adsorbent. Adsorption is caused by surface forces acting on the interface between a solid and a gas or between a solid and a liquid or (in the event of immiscible liquids in contact) on the interface between a liquid and a solid.

There is no distinction between physical adsorption and chemisorption when it comes to the adsorption of gas by a solid surface. In physical adsorption, the nature of the interacting substances, the forces of contact at the interface, and the interaction between adsorbate molecules are the most important aspects to take into consideration.

The quantity of surface inhomogeneity present, as well as the capacity of adsorbed molecules to move in translational, rotational, and oscillatory directions, are all critical considerations. During the adsorption process, the surface structure and, consequently, the adsorption characteristics may vary. In particular, when the so-called adsorption energy of the adsorbent is on the same order as the surface energy of the adsorbent, as in the case of ice, paraffin, and polymers, the effect is most noticeable.

Afforestation

An area that has not previously been wooded is being reforested by the introduction of trees and tree seedlings. Afforestation can be accomplished by the planting and seeding of trees, either naturally or intentionally induced. As with afforestation, reforestation may be regarded a type of afforestation in its own right.

Reforestation is the process of converting a non-wooded region into a forested one by the planting of trees and the sowing of seeds. What makes reforestation distinct from other forms of forest regeneration is that it involves the restoration of previously deforested areas. There are a variety of reasons for recovering a region through afforestation and reforestation, and the reasons vary depending on the location. In most cases, it is done for conservation or for industrial-commercial objectives, rather than both.

Forest conservation and replanting are carried out in the best interests of the environment in which they are carried out. In order to repair an area that has been devastated due to prior abuse of the land, or to minimize the quantity of soil erosion in an area, it is necessary to create a more productive and solid soil base. It is necessary to do industrial-commercial afforestation and reforestation in order to maintain a sufficient supply of wood for pulp and timber requirements in a given region.

Agglomerate

Agglomerate is a term used to describe massive, coarse rock particles that are thrown along with lava flows during explosive volcanic eruptions. Despite their resemblance to sedimentary conglomerates, agglomerates are pyroclastic igneous rocks that are nearly entirely composed of angular or rounded lava pieces of varied size and form, as opposed to sedimentary conglomerates. When found in a tuffaceous matrix, fragments are often poorly sorted, or they appear as lithified volcanic dust or ash.

Some geologists prefer to categorize agglomerates into three categories: bombs, blocks, and breccia, depending on the unique setting. Generally speaking, bombs and blocks are bigger than 32 mm (1.25 inches) in size; although bombs are expelled while still in a molten condition (and become rounded upon solidification), blocks are released as solid angular or subangular chunks after they are solid. Breccia, which are solid angular fragments greater than 64 mm in size, are formed as a result of the accumulation of blocks.

Aggressivity

Aggressive CO_2 and other metrics of aggressivity, such as the Langelier Saturation Index (LSI) and the Aggressive Index (AI), are only heuristics. In order to attempt to express anything meaningful about the water for practical purposes such as specifying building materials or sizing acid or alkali dosing pumps, they make use of some of the factors.

They are all connected to one another, but their interrelationships are complicated since each heuristic stresses a different element than the others.

Agricultural Geography

The study of geographical and locational attributes, patterns, and processes of crop and animal farming, as well as related subjects such as farmland, farm-associated human geographers, environmental issues, and theoretical works on agricultural activity location is known as agricultural geography. Agricultural geography is divided into two categories: theoretical works on agricultural activity location and applied studies on agricultural activity location.

The study of agricultural geography has resulted in a substantial volume of written material. Attempts have been made to record and present as much of this material as is feasible in an orderly fashion. The items in this compendium are written in a diverse range of languages, including English, French, Spanish, Portuguese, Italian, German, Swedish, Danish, Dutch, and Russian, among others, in order to give the broadest possible coverage of the subject matter covered by the compendium.

Atlases, novels, book chapters, academic papers from professional journals, conference proceedings, doctorate dissertations, and master's these are among the items included in the database. In the expectation that such references would inspire and support the work of students, teachers, and other users, more than 12,000 entries have been logged here.

Agricultural Revolution

In the 18th century, British started a radical transition of its old agricultural system, which became known as the Agricultural Revolution. This complex transformation, which was not completed until the nineteenth century, included the reallocation of land ownership to make farms more compact, as well as increased investment in technical improvements such as new machinery, better drainage, scientific methods of breeding, and experimentation with new crops and crop rotation systems, all of which were completed by the nineteenth century.

For example, the Norfolk four-course system, which was developed in Norfolk County, England, placed an emphasis on fodder crops and did away with the fallow year that had previously been used as a standard crop rotation technique. Wheat was planted in the first year, turnips in the second, barley in the third, and clover and ryegrass were seeded under cover in the fourth and final year.

In the fourth year, the clover and ryegrass were harvested for fodder or grazed for pasture. Turnips were given to cattle and sheep throughout the winter months. Scientific breeding enabled the creation of Shorthorn beef cattle by selective breeding of native cattle in Durham county's Tea water district, which was representative of the advancements made possible by scientific breeding.

A lot of people have forgotten about the agricultural revolution and how it was a prelude to the Industrial Revolution, but the names Jethro Tull and Arthur Young are still frequently mentioned by those who want to understand the significance of this period in history and how it was influenced by the contributions of great men.

Agricultural Systems

There are several ways in which agricultural systems might be defined. Over the years, academics and farmers alike have used a number of terminologies to describe agricultural systems based on specific features or definitions, such as farming system, cropping system, organic, and ecological, to identify agricultural systems. Other definitions for example, a corn–soybean system, a vegetable or hog production system are based on the commodity being produced, as opposed to these words, which are based on specific sets of practices, management approaches, and occasionally ideologies. The word agricultural system is referred to any system that produces animals and crops such as food and feed, fiber and energy, as well as any system that has social, political, and economic components.

Agricultural systems researchers should not feel constrained by these criteria when conceiving their own systems for research and instructional purposes. In order to effectively build or designate their own systems, people should select and choose from a variety of options.

B

Backshore

The backshore is that part or zone of a sea shore profile that broadens landward from the slanting foreshore to a state of either vegetation advancement or a difference in geography, for example, an ocean bluff or a hill field.

There might be an unmistakable irregularity to the spatial meaning of the backshore, because of summer to winter changes in elevated tide levels just as tempest designs. In these cases, the botanical and faunal arrangement of the backshore may fluctuate among seasons.

Albeit the biodiversity of the backshore is a lot higher than that of the requesting climate of the foreshore, there are a lot more modest number of species that possess backshores contrasted with more upland environments. This result streams from the interestingness of life in the desolate living space where the dry sands are always moving, and saline soils are predominant.

Backswamp

The distal floodplain, at valley edges, is regularly the lowest space of the valley floor. They are significant capacity units of fine-grained, in an upward direction accumulated, suspended burden residue. Morphology is normally genuinely level (or has low alleviation), with despondencies. Lakes, wetlands and bogs generally structure where lower request feeders channel straightforwardly onto the floodplain.

Structures when the decrease in energy slope from the proximal to distal floodplain just permits suspended burden materials to be moved to the backswamp. This outcomes in lethargic paces of fine-grained vertical growth in these settings. An unmistakable degree in energy with distance from the divert may result in articulated textural isolation across the floodplain.

Backswamps, wetlands, lakes and lake highlights are normal in these inadequately depleted (unchanneled), low-energy, in an upward direction accumulating conditions. Normally colonized by thick sea-going/swamp vegetation that traps fine-grained suspended-load silt. These advances gathering of firm, natural rich mud. Materials are commonly profoundly bioturbated.

Balanced Growth

Balanced growth has at any rate two distinct implications in economics. In macroeconomics, adjusted development happens when yield and the capital stock develop at a similar rate. This development way can excuse the since quite a while ago run solidness of genuine loan fees, yet its reality requires solid suppositions. Being developed financial matters, adjusted development alludes to the synchronous, composed extension of a few areas. The standard contentions for this advancement system depend on scale economies, with the goal that the efficiency and productivity of individual firms may rely upon market size. The article audits the reasonable development banter and the degree to which it has affected advancement arrangements.

Bank Erosion

Bank erosion is second key interaction in fluvial elements, which affects a progression of physical, environmental, and financial issues. It has become dynamically perceived that some sidelong elements related with bank disintegration have beneficial outcomes as far as the advancement of riparian vegetation progression, the foundation and development of stream and floodplain morphology, and the production of dynamic territories pivotal for amphibian and riparian species (Florsheim et al., 2008).

Figure 4. River bank erosion.

Source: Image by Wikimedia.

The recognizable proof and forecast of the spatial dissemination of bank measures, the propensity to sidelong channel portability, and its controlling variables aggregately structure a significant issue. As a first estimation, the sidelong portability dispersion at a waterway network scale can be identified with cooperation between Stream Energy (or force) and limit opposition. It is likewise very much perceived that bank retreat is the coordinated result of three collaborating gatherings of cycles (subaerial measures, disintegration cycles, and mass squandering).

In blend, these recommend the presence of changing bank measure predominance across a speculative seepage bowl in light of downstream varieties in energy conditions, bank material, and climatic conditions (see, e.g., Lawler, 1992). All the more as of late, Finstad and Marcus (2003) have speculated the presence of self-coordinated in riverbank frameworks, to such an extent that they can change inside (i.e., put together themselves) with no adjustment of the greatness and recurrence of outside contributions to the framework.

Backreef

Three sites in the red sea were researched to evaluate the fluctuation of organization in Holocene residue of the backreef climate inside 0–2 m of water profundity. This is significant in light of the fact that creation of the silt is ordinarily used to appraise water profundity in antiquated carbonate rocks. The site situated at the king Abdullah economic city (Saudi Arabia) contains a bordering reef with the reef plot found near the sea shore at the north end, erupting toward the south to create a smaller backreef region contrasted with the other two destinations.

This calculation delivers a north to south current with a speed of up to 15 cm s^{-1}, especially during high coastal breezes. The silt contain prevalently non-skeletal grains, including poloids, covered grains, ooids, and grape stones that structure on the base. The level of coral gal grains in the residue was fundamentally lower than at the other two locales considered. Om Al Misk island and Shuaiba have a much lower-speed current inside the backreef zone and contain dominatingly coral gal silt from the sea shore to the landward edge of the reef parcel.

The two areas containing the overwhelmingly coral gal microfacies were genuinely comparable, yet the king Abdullah economic city site was measurably extraordinary in spite of having a comparable water profundity profile. Slight contrasts in reef setup, including reef direction and distance from the shore, can deliver impressive contrasts in silt thickness and organization inside the backreef climate, which ought to initiate alert in the understanding of water profundity in old carbonate rocks utilizing arrangement.

Banner Clouds

Banner clouds are accepted to shape when wind streams around a mountain (instead of over it). Vortices on the two sides of the mountain are shed to the back and produce upslope stream, inverse in heading to the common breeze, on the lee side of the mountain. This upslope stream is attracted to a non-hydrostatic pressing factor least beneath the mountain peak.

This low is an overall vacuum because of the stream decrease behind the hindrance (as indicated by the Bernoulli hypothesis). A similar stream example can likewise be clarified through protection of expected vorticity. The upslope stream may arrive at the lifting build-up level, causing shadiness. The overcast air in the lee of the mountain is ceaselessly blended in with encompassing air and shed to the back, clarifying the banner like appearance.

Figure 5. Banner clouds.

Source: Image by flickr.com.

Barchan Dunes

Barchan dunes address the exemplary dunes morphology, showing a crescentic shape with peaks adjusted opposite to the overall breeze heading and horns guided corresponding to it.

These dunes types range in size from at least 1 m in tallness (Elbelrhiti et al., 2005) to more than 50 m and, existing in districts with little wind directional fluctuation (ordinarily close to beach front areas or on the edges of ridge fields), they are the most unique as far as speed of downwind development and together make up around 10% of desert rises worldwide (breed and grow, 1979). Barchans and cross over edges address transitory structures and are the kinds of hill that cause the most risk to human action concerning accumulation on streets, railroads, and correspondence lines (Dong et al., 2000, 2004b).

Estimations by Hesp and Hastings (1988) uncovered that barchan dunes have a genuinely restricted reach in illustrative measurements with the most extreme stature being of the request for 1/tenth of both the length (upwind toe to verge) and horn-to-horn width. Windward inclines of barchans are for the most part of the request for 10–14° with downwind slip-faces at the point of rest of sand (32–34°). Numerous varieties as barchan dunes happen with some showing fortuitous event in the area of peak (most noteworthy point) and verge (most elevated mark of the slip-face) and others showing differing levels of peak edge division

Batholith

An exceptionally enormous, arch like mass of barged in igneous rock. At first, the interruption happens at an extraordinary profundity inside the earth's hull, in relationship with earth developments and mountain building. Notwithstanding, after an extensive stretch of denudation, the upper surfaces of the batholiths are in the long run uncovered at the surface.

Bedding Plane

The surface isolating individual layers of a sedimentary rock like limestone, chalk or sandstone. Bedding planes frequently appear as breaks along which underground water can move in a down-dip heading. They additionally establish lines of shortcoming that can be misused by weathering measures (for instance, freeze-thaw weathering), so that very much slept with rocks will in general be moderately open.

Figure 6. Ceiling bedding plane.

Source: Image by flickr.com.

Biomass

The absolute natural matter contained by plants as well as creatures, typically communicated as far as oven-dry weight per unit area. Plants, both living and rotting, comprise by a wide margin the best extent of the earth's biomass. Creature biomass (zoo biomass) is little by correlation and quite a bit of it comprises of miniature organic entities in the dirt. Plant biomass differs extensively with environment and vegetation type. In the prairie fields of America, most extreme biomass (at the tallness of the developing season) is around 10,000 kg ha^{-1}; in the tropical savanna it is 60,000 kg ha^{-1}; and in deciduous oak forest it is 250,000 kg ha^{-1}.

Figure 7. Wood biomass.

Source: Image by Pixabay.

Biome

A world-scale ecosystem generally characterized by its prevailing type of vegetation, and ordinarily connected with a specific environment. Tropical meadow and savanna is one illustration of a biome. The forests and lush fields of Africa (e.g., the Serengeti plains), with their huge groups of perusing and touching creatures (elephant, giraffe, wildebeest, zebra and gazelle), and their related hunters (lion, panther, cheetah, hyena and jackal), compare to a semi-dry heat and humidity. Different instances of significant world biomes incorporate tropical rainforest, tundra, and Mediterranean forest and scour.

Biotechnology

A term applied to a wide scope of exercises joined by the way that each tackle at least one of the unique capacities of living cells. It is at present a space of impressive R&D coordinated towards such explicit fields as genetic engineering, microbial mining, and the revelation of new anti-toxins and immunizations. It is right now delegated one of the main high-technology industries.

Figure 8. Biotechnology plays a key role in today's science and technology.

Source: Image by Pixabay.

Block Disintegration

The breakdown of rock into huge squares by both mechanical weathering and chemical weathering. The cycle relies upon the presence of lines of shortcoming (gaps, joints and bedding planes) that can be entered by weathering specialists, especially acidic water. One significant type of square breaking down outcomes from the freezing of water that has entered breaks in the stone.

The ensuing development by 9% in volume as the water sticks to frame ice causes the wedging separated of breaks and, in the end, the actual crumbling of the stone. In any case, in warm, sticky environments corrosive water can, by cycles like hydrolysis, open up joints in rocks like granite, and again lead to hinder breaking down. In this example the subsequent squares for the most part show proof of adjusting, while a simply actual cycle, for example, ice wedging produces forcefully rakish, joint-limited squares.

Boreal Forest

The generally coniferous forests possessing immense spaces of America and Eurasia principally between the scopes 45°N and 75°N, offering a genuine illustration of a biome. Climatic conditions here are cruel, with cold winters and brief summers. There is a short developing season, consistently of under a half year and at times of just 3 months, and precipitation is low (up to 500 mm yr^{-1}) however satisfactory for tree development. Over a significant part of the boreal woods zone the trees are evergreen; species like fir, pine and tidy are prevailing. These are adjusted to the natural conditions (e.g., the short and adaptable branches shed substantial snow, and the little needle-like leaves lessen happening during winter.

Figure 9. Winter boreal forest.

Source: Image by Wikimedia.

Bottom-Up Development

A hypothesis of development that calls for assets and ventures to be focused on the provincial poor with an end goal to lessen destitution and meet fundamental human requirements. It is the inverse to stream down or hierarchical turn of events, which was the primary hypothesis behind unfamiliar aid until the last part of the 1960s.

Boulder

An enormous piece of rock with a width more than 200 mm. Stones are at first disconnected from the bedrock by an interaction like block disintegration and are then changed by transportation measures, getting diminished in size and less rakish.

Boulder Clay

An unstratified mass of ineffectively arranged glacial deposits containing huge, frequently rakish stones and even rocks, held together by a 'clayey' matrix. Stone earth frames a shallow store that covers numerous spaces of swamp glacial deposition, and may accomplish a thickness of many meters. The store includes chiefly ground moraine framed by the basal softening of debris-rich glacial. Running water, nonetheless, has no impact in the statement of stone earth, consequently its unstratified nature and the strength of angular stone particles. Rock mud is a frail and unconsolidated store that is promptly dissolved, especially when uncovered at the coast.

Boundary

The splitting line between one political state and another. All the more broadly utilized in geology to signify the division between discrete elements (e.g., land outcrops, climatic sorts, financial districts, social regions). Limited levelheadedness an idea of behavioral geography set forward as a response to the set-up perspective on the leader as a normal financial individual.

All things considered; it is contended that:

i. the data on which dynamic is based isn't openly accessible, yet is compelled by time, monetary assets and perception;

ii. chiefs have a restricted ability to handle any such data that they secure; although leaders may endeavor to act soundly, they will unavoidably be obliged (limited) by their own behavioral environment and that, correspondingly, they will be substance to receive imperfect arrangements.

Braided Stream

A stream in which there isn't one single channel, yet a progression of little interconnecting channels (some utilized constantly, some utilized uniquely under states of high discharge) isolated by little bars or bigger, stable and vegetated islands. Twisted streams happen:

i. in regions where the channel banks are effectively disintegrated (for instance, where they are made out of free sands and gravels);

ii. where the release is profoundly unpredictable (as in cold meltwater streams, which experience both occasional and diurnal stream varieties). Interlaced stream channels are powerfully wasteful, and are described by steep longitudinal inclinations. These are important to elevate the speed expected to move the water and bed load through the various channels.

Branch Plant

A subordinate or auxiliary division of a business, normally settled to fulfill an expanding need, and frequently found away from the parent organization yet close to the new market or some modest resource. For instance, Japanese firms have set up branch plants in Britain to build their deals of items like vehicles and family merchandise.

Not exclusively are their cost benefits to be acquired by the firm assembling in Britain as opposed to transportation the completed merchandise there from Japan (benefits that are expanded even further by the accessibility of monetary assistance should the branch plant be situated in a development area), however deals of Japanese products made in Britain are likewise a substitute for direct imports from Japan. Accordingly, consequently, they help to diminish the embarrassingly enormous exchanging surplus that Japan has with the UK—for example, they have diminished trade friction.

Break-of-Bulk Point

This is the place where payload is moved starting with one method of transport then onto the next, as at a port or railroad station. Such focuses are critical as far as financial area in that they offer possible reserve funds in transport costs. It tends to be seen that by preparing raw materials at the break-of-mass point, investment funds are made for the accompanying reasons: there is no exchange of crude materials from boat to rail, so no transshipment costs are brought about; there is no deficiency of advantage because of tightening freight rates. The meaning of break-of-mass focuses as most minimal vehicle cost areas was pushed in Hoover's theory of the location of economic activity.

Break of Slope

A plainly noticeable, sharp difference in steepness in a slant profile or waterway long-profile. Breaks of slant regularly result from topographical impacts (a change from hard to delicate rocks) and the systems of incline downturn (for instance, the quick retreat of a precarious precipice portion 'devouring' a gentler slant component above).

Break-Point Bar

A sand bar framed, on banks of exceptionally shallow inclination, at the line of wave break. They are believed to be shaped by the breaking of steep waves that cause an offshore float of sand inside the break-point, where the resultant gathering structures the bar.

Brownfield Site

Land that has been utilized, deserted (derelict land) and now anticipates some new use. Generally found in urban regions, especially in the downtown where land has been made empty by plant terminations, and the decentralization of individuals and exercises to the suburbs and past. Brownfield destinations are usually involved by neglected structures and supported for unloading junk. Thusly, they are awful for the picture of downtown. It is contended that redeveloping these destinations ought to have need over the improvement of greenfield sites in the open country. The fact is made that such locales are less expensive and extra the open country, and that their utilization will help resuscitate downtown regions.

C

Capacity

In nature, conveying limit is estimated as the most extreme heap of a climate. The actual highlights present in the climate go about as restricting elements (e.g., food, water, rivalry, and so on) subsequently, as far as possible can be relied upon to rely upon these variables. Basically, food accessibility is a significant variable as it influences the populace size of the species.

It does as such so that if food request isn't met for throughout a given timeframe the populace size will at last diminish until the assets become sufficient. Paradoxically, when food supply surpasses request then the populace size will before long increment and will quit expanding when the source is thus drained.

Conveying limit may likewise be characterized as the populace size at which the populace development rate approaches zero. It ought not be mistaken for the term, balance populace, which is characterized as a populace in which the quality frequencies have arrived at a harmony between change pressing factor and determination pressure.

Calcrete

Calcrete, additionally called hardpan, calcium-rich duricrust, a solidified layer in or on a dirt. It is framed on calcareous materials because of climatic variances in bone-dry and semiarid locales. Calcite is broken down in groundwater and, under drying conditions, is encouraged as the water dissipates at the surface. Water immersed with carbon dioxide goes about as a corrosive and furthermore disintegrates calcite and afterward redeposits it's anything but an accelerate on the surfaces of the dirt particles; as the interstitial soil spaces are filled, an impermeable outside is framed.

Capital Intensive

Capital escalated alludes to a useful cycle that requires a high level of interest in fixed resources (machines, capital, plant) to deliver. A capital-concentrated creation cycle will have a generally low proportion of work inputs and will have higher work usefulness (yield per specialist).

A capital concentrated creation cycle will in general have a high proportion of fixed expenses for variable expenses. Thus, the useful cycle will have economies of scale (expanded yield prompts below costs). Lately, innovative advancement have empowered expanded capital force in numerous enterprises. New innovation like artificial intelligence, miniature pcs and the web have empowered already work concentrated enterprises (taxi's, conveyance to turn out to be more capital serious – for example, with the ascent of self-driving vehicles.

Expanded capital force can make a few laborers lose their positions since they are not, at this point required. This can prompt underlying joblessness – at any rate temporarily. Be that as it may, more capital escalated enterprises make various types of occupations. There are new positions in the plan of programming, ai and promoting. Additionally, capital-serious creation can prompt lower costs and higher livelihoods. This causes expanded interest for a more prominent assortment of administrations in the economy.

Capital Labor Ratio

In communist financial aspects, a pointer that describes the amount of fixed creation resources in parts of material creation on a for every laborer premise. The proportion is gotten by separating the book worth of these resources for a given year by the quantity of laborers utilized during that year. Stipends are made in the sums for laborers (and resources) utilized for just piece of the year.

The capital work proportion is utilized in financial examination and arranging. Proportions shift from one industry to another attributable to contrasts underway cycles and in designing and configuration highlights and costs of fixed capital stock. In 1975, the capital-work proportion for industry all-in-all the USSR was roughly 9,000 rubles; on Kolkhozes, Sovkhozes, and other state farming ventures, it was 6,200 rubles. Contrasts in the proportion are significantly more articulated between the different parts of industry. Around the same time, there were 3,000 rubles of capital for every laborer in light industry, 22,000 rubles for each specialist in ferrous metallurgy, and just about 96,000 rubles for each laborer in the oil-extraction industry.

The ascent of the capital-work proportion is a target financial cycle achieved by the quantitative and subjective development of the methods for work. This ascent, which is normal for all

Capitalism

Practically talking, private enterprise is one interaction by which the issues of monetary creation and asset conveyance may be settled. Rather than arranging monetary choices through unified political techniques, likewise with communism or feudalism, financial arranging under private enterprise happens by means of decentralized and deliberate choices.

- Capitalism is a financial framework described by private responsibility for methods for creation, particularly in the modern area.
- Capitalism relies upon the requirement of private property rights, which give impetuses to interest in and useful utilization of useful capital.
- Capitalism grew verifiably out of past frameworks of feudalism and mercantilism in Europe, and significantly extended industrialization and the huge scope accessibility of mass-market customer merchandise.
- Pure free enterprise can be stood out from unadulterated communism (where all methods for creation are group or state-claimed) and blended economies (which lie on a continuum between unadulterated private enterprise and unadulterated communism).
- The genuine act of free enterprise ordinarily includes some level of alleged "comrade private enterprise" because of requests from business for ideal government mediation and governments' motivator to intercede in the economy.

Carbon Cycle

Carbon is a fundamental component to all living things on earth – plants and creatures. It's anything but a significant job in controlling worldwide environment, especially temperature and in characterizing the acridity of downpour, streams and seas. The carbon cycle includes the capacity and development of carbon. The carbon cycle is overwhelmed via land and ocean stores. Carbon quickly circles the environment, soil, seas and biosphere.

There are two components to the carbon cycle. One is the lethargic carbon cycle, the other, the quick. The long carbon cycle includes the drawn-out capacity of carbon. Marine creatures, like shellfish and phytoplankton, assemble their shells by consolidating calcium with carbon. At the point when they bite the dust, they gather on the sea floor. More than a long period of time, these organic entities become packed and become carbon-rich sedimentary stone. This carbon is typically put away in rocks for around 150 million years.

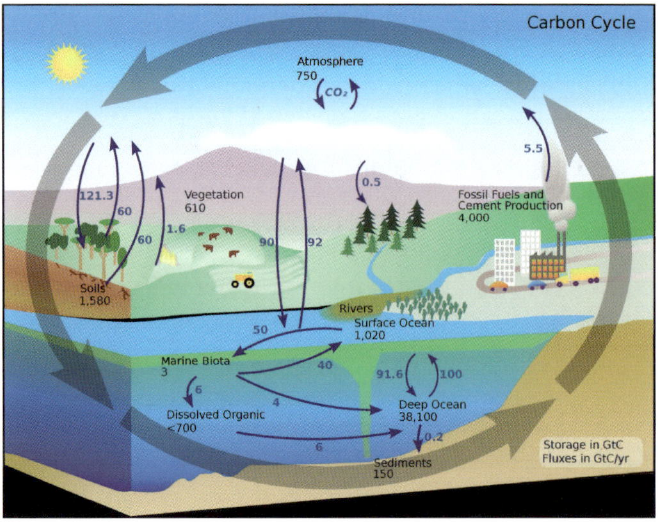

Figure 10. Carbon cycle.

Source: Image by flickr.com.

Carbonation

Carbonation is the immersion of a fluid with CO_2 gas. All in all, it's anything but a term used to depict the disintegration of CO_2 gas in water using pressing factor and temperature. It ordinarily incorporates cold CO_2 under high tension. CO_2 is a nontoxic, idle gas that is essentially dull and is promptly accessible at reasonable expense. It is dissolvable in fluids and can exist in the three matter stages, specifically as a strong, fluid, or gas.

This cycle can happen normally or by fake cycles, similar to the case in most delicate carbonated beverages and soft drink water. The most extreme measure of CO_2 that can be disintegrated in water is 8 g/l. The overabundance of CO_2 will regularly possibly stay in water when the beverage is feeling the squeeze. Or on the other hand as such, the CBS are set up by blending chilled seasoned syrups in with carbonated water in which carbonation levels range up to 3.5–5 g CO_2 for every fluid volume in colas and related beverages, while fruity ones, are less carbonated.

Carnivore

Carnivore, likewise called hunter, creature whose diet comprises of different creatures. Transformations for a flesh-eating diet incorporate an assortment of chasing practices and the advancement of techniques for getting a handle on or in any case immobilizing the prey. Wolves utilize their teeth for getting a handle on, owls their paws, and bullfrogs their tongues. A few snakes (e.g., diamondbacks) use toxin to immobilize their prey, and numerous arachnids envelop their casualties by string. Most carnivores are bigger than their prey species, albeit probably the biggest carnivores go after much bigger species—for instance, tigers on water bison and orcas on baleen whales.

Some notable hunters (e.g., lions, weasels, canines) have a place with the mammalian request carnivora. In any case, savage warm-blooded animals are additionally found in numerous different orders, including the insectivore, Cetacea, marsupial, and chiropteran. Similarly, not all vertebrates in the order carnivora are solely carnivores. Certain plants are additionally flesh eating. The Venus flytrap has leaves altered to go about as snap traps. Different plants are outfitted with latent gadgets that work like flypaper or lobster pots.

Figure 11. Lion is an example of carnivore.

Source: Image by Britannica.

Cartography

Cartography, the workmanship and study of graphically addressing a topographical region, for the most part on a level surface like a guide or outline. It might include the superimposition of political, social, or other nongeographical divisions onto the portrayal of a topographical region.

Cartography is an antiquated control that dates from the ancient portrayal of chasing and fishing domains. The Babylonians planned the world in a leveled, plate molded structure, yet Claudius Ptolemais (Ptolemy) set up the reason for resulting endeavors in the second century.

Guides created during the Middle Ages followed Ptolemy's aide, however they utilized Jerusalem as the focal component and set east at the top. Those portrayals are frequently called t-maps since they show just three landmasses (Europe, Asia, and Africa), isolated by the "t" shaped by the Mediterranean Sea and the Nile River. More exact geological portrayal started in the fourteenth century when portolan (seamen's) graphs were incorporated for route.

Figure 12. Compass map cartography.

Source: Image by Pixabay.

Cascade

The word course invokes a dream of anything falling in a way taking after a waterfall—think about a course of sparkling hair, an unexpected course of tears, or the course of light during a meteor shower. The thought behind the word is of something falling in a reformist way or a progression of steps. A descending movement of melodic subjects can be known as a course, as can the terrible course of the tumbling cost of a stock on wall street.

Cash Crop

Money crop requires broad administration of hazard. The harvests being a fundamental piece of practical strengthening as pay created with cash crops give ranch families. The pay assists them with saving and put resources into a more useful homestead. Money yields may catalytically affect agribusiness advancements as they add worth and usefulness in country regions.

A horticulture crop developed to get benefit is alluded to as a money crop. A money crop is exclusively created to fare or selling instead of for utilization by the individual who develops it. Financial experts utilize the term cash yields to allude to any edit that is effectively sold in the open market, like sugarcane, cotton, or tobacco. It is the harvest that verifiably has delivered a high pace of profit from the cultivator's venture.

Census

The assortment of segment, monetary and social data pretty much all or an example of individuals in a characterized region at a specific time. Public censuses have been attempted since the eighteenth century. To be of genuine worth, censuses should be taken at customary stretches (when at regular intervals is a generally noticed standard) and to gather similar kinds of information for similar arrangement of spatial units. By noticing these three standards, the aftereffects of progressive censuses can measure up and patterns of progress all the more promptly recognized.

As respects the subsequent rule, the un has suggested that all public censuses ought to specify the accompanying: complete populace; age, sex and conjugal status; spot of birth and ethnicity; proficiency and instructive accomplishment; family and family structure; richness; country or metropolitan home. There are two distinctive registration techniques. The accepted methodology records populace where it is figured out at the exact season of the registration (e.g., as in Britain), while the by law approach includes recording individuals as indicated by their standard spot of home.

Central Businesses

Focal place functions usually found in the focal region of a town or city, especially the retailing of higher-request products, proficient administrations (monetary, lawful, clinical), individual administrations (stylists, dry-cleaners, travel planners) cooking and amusement, wholesaling and specific sorts of expert industry (e.g., printing and distributing, style clothing).

A few geographers would limit the term to incorporate just benefit making exercises (as above), while others take a more extensive view and would incorporate exercises falling under the expansive heading of public utilities, like neighborhood government workplaces, libraries and galleries, schools and universities, sports focuses, crisis administrations, and so forth focal organizations normally require a focal area to expand admittance to clients drawn from all pieces of the developed region and from the encompassing market area, and the get together of work for what are distinctively labor-intensive exercises.

A few organizations are attracted to the focal city since they see it's anything but a renowned area. Outside economies additionally have an influence in the centralization of most sorts of focal business. As a rule, undeniable degrees of turnover and enormous net revenues empower focal organizations to bear the cost of the great land esteems and leases that win in such focal and open areas.

Central Eruption

An emission of volcanic lava, soot or ash from a solitary vent or gathering of firmly dispersed vents coming about for the most part in the development of a volcanic cone. Inside a space, various focal ejections may happen along a fault – line (or arrangement of shortcomings), leading to a line of volcanic cones. Most exemplary volcanoes, like mt Pinatubo in the Philippines, are instances of focal vent volcanoes.

Central Place

Any settlement giving labor and products to the advantage of an encompassing feeder region (market area) that may contain both rural regions and more modest, subordinate settlements. While most towns, towns and urban communities work as focal spots, appreciate that as the size of settlement increments, so singular settlements foster their own interior frameworks of focal spots. For instance, the traffic intersection store and the rural shopping precinct may be viewed as making up various orders of focal spot inside the built-up area of a town or city.

Chemical Weathering

The decay of rock minerals in situ (in their unique situation) by specialists like water, oxygen, carbon dioxide and natural acids, where a compound change happens. Among the main individual cycles of substance enduring are carbonation, chelation, hydrolysis, oxidation and solution. Compound enduring is at its best in muggy heat and humidity where synthetic responses are supported by high temperatures plentiful soil moisture, and the age of humid acids by rotting vegetation.

In any case, it is presently known to be dynamic even in chilly arctic environments, where carbon dioxide gets gathered in snow banks, and natural acids are related with lowland vegetation (carbon dioxide is more solvent in cool conditions than warm). Albeit an unmistakable differentiation is regularly made among compound and physical weathering, the two kinds much of the time cooperate in causing the separation of rock.

Figure 13. Chemical weathering.

Source: Image by flickr.com.

Containerization

The development of merchandise in huge normalized metal boxes that are fit for being conveyed by three unique methods of transport (boat, street and rail). Containerization of freight implies a simpler and more effective exchange of merchandise between transport modes, just as empowering products development to profit with the specific economies related with those modes. In this manner street transport gives a house-to-house administration taking care of to and from railroad warehouses; rail works a fast help moving the compartments to and from port terminals; while ocean transport gives modest, long stretch carriage to other import or fare terminals

Containerization has implied decrease both in the quantity of individuals utilized in cargo handling exercises, since these have gotten exceptionally motorized, and in the meaning of transshipment focuses as modern areas. It has additionally implied the advancement of new and enormous compartment conveying ships, just as the establishment of broad and exceptionally robotized transport terminals.

D

Dead Ice

Ice, framing part of an icy mass or ice sheet, that is done streaming. Dead ice is found in an assortment of circumstances. For instance, when an alpine glacier goes through flooding because of expanded precipitation in the accumulation zone, the nose will progress impressively. In any case, when the driving force of the flood has finished, the lowermost piece of the ice sheet will be deserted as a dead ice mass, gradually dissolving endlessly in situ and getting covered by ablation moraine. At the point when dead ice liquefies, it typically leaves behind hummocky ground, regularly with highlights of glacio-fluvial testimonies, for example, kames and kettle holes.

Figure 14. Dead ice over the stream.

Source: Image by Imaggeo.com.

Death Rate

The normal number of passing per 1000 occupants in a given populace. It's anything but an extremely refined proportion of mortality, since it doesn't consider the general age-sex design of the populace. A high passing rate in a 'youthful' populace is an essentially extraordinary circumstance from that where a similar rate wins in a 'more established' populace.

Debris

Dissipated and broken pieces regularly coming about because of the rot and crumbling of rocks by enduring cycles. Such material involving slants is named scree or talus.

Debt

In topography this alludes mostly to what in particular is frequently called third world obligation. It is obligation owed by led to worldwide aid associations. Third world obligation expanded hugely during the last quarter of the twentieth century. The base of the issue lies in the way that amount of the guide was 'given' as credits. In addition to the fact that loans have to be reimbursed, however they likewise draw in interest – this too needs to possibly be paid.

Obligation reimbursement is just conceivable if the economy of the getting nation thrives. Shockingly, this has once in a while occurred. This, in addition to increasing financing costs, has implied nations falling increasingly more behind in their reimbursements. Giving further advances to take care of obligation back payments just aggravates the emergency as nations essentially fall further into obligation (presently alluded to as the obligation trouble).

They enter a descending endless loop known as the obligation twisting. Obligation expands dependence on the benefactor country; it makes the rich more extravagant and the poor more unfortunate, along these lines broadening the development gap.

As of late, some MEDC have concurred on a fundamental level to obligation alleviation, either to clear or pay off the worldwide obligations of the most unfortunate nations. At this point, nonetheless, there are not many signs that anything has been finished. Meanwhile, a considerable lot of the worlds LEDC (Less Economic Developed Country) are caught as survivors of this 'new bondage' made by global obligation.

Decalcification

A dirt shaping cycle, including the expulsion of calcium minerals by penetrating water. It influences soils in regions where precipitation is more noteworthy than vanishing over a large part of the year. In sticky mild timberland areas decalcification and humifaction (the joining of humus inside the dirt) bring about earthy colored earth.

Decentralization

The activity of centrifugal forces prompting outward development (e.g., of individuals, financial exercises) from set up focuses. This is shown in the core-periphery model, which shows a trickle down of development from center to outskirts.

It's anything but an essential piece of suburbanization and counter urbanization, and includes 'getting' populace and work from the internal spaces of urban areas and 'putting down' either in the suburbs or in more modest urban focuses and rural regions.

Such developments are regularly an intentional reaction to the negative externalities of enormous urban communities (particularly of their more seasoned regions) and the positive externalities that are seen in the new areas. Arranged decentralization from the inner city was important for the overspill program in the UK during the 1950s and 1960s.

Deciduous Forest

A kind of backwoods including trees that experience yearly leaf fall (typically in the harvest time or 'fall').

These trees are fundamentally expansive leafed (like oak, beech, debris, birch and maple), however a few conifers in center and high scopes (e.g., larch) have the deciduous propensity. Deciduous trees are preferred adjusted over conifers to contending in the climatic states of regions like Europe.

Specifically, they are balanced to the more extended developing season, with a half year or a greater number of mean temperatures in overabundance of 6°C, and to the yearly well distributed precipitation (in the reach 750–1000 mm), which gives a colder time of year excess repaying to any late spring shortfall brought about by evapotranspiration surpassing precipitation.

The shelter of the woods gives just fractional shade in summer (and next to no in winter and spring), with the goal that an undergrowth of saplings, bushes, blossoms and grasses can flourish.

The soils underneath summer deciduous backwoods trees are more extravagant than those of the boreal forest attributable to yearly additions of leaves providing reflect on humus. Most 'normal' deciduous timberland has been obliterated or adjusted by individuals.

Figure 15. Deciduous forest.

Source: Image by Pixabay.com.

Decision Making

The interaction whereby elective game-plans are assessed and a choice taken. A significant component of examinations of dynamic is the acknowledgment that area choices are only from time to time, if at any time, ideal in the feeling of expanding benefits and limiting assets utilized. Likewise, purchaser conduct and dynamic barely at any point accord with the suppositions made, for instance, in central place theory.

Along these lines, two elective ideas safe house set forward: the satisficed concept and the idea of suboptimal location. Endeavors to make speculations regarding dynamic are steadily baffled by inevitable realities: the huge variety of chiefs; their fluctuating forces of perception; their various conditions; their shifting admittance to significant data; contrasts in their capacity to deal with that data and to assess elective approaches.

Decomposers

Living beings (especially microorganisms and growths) that feed on plant and creature remains, separating them and in the process delivering energy and supplements. Decomposers assume a significant part in keeping up soil richness, by helping the rot of leaves and other plant material at the dirt surface, and the consolidation of the resultant items inside the dirt.

Deep Weathering

The creation of a thick regolith (once in a while alluded to as saprolite) by delayed and additionally serious chemical weathering. Profound enduring is particularly connected with spaces of low relief in sticky tropical conditions, where warmth and stickiness favor cycles like hydrolysis, oxidation and carbonation, yet where transport is limited by the delicate slants and restricting impact of the thick vegetation. It has been proposed that profound enduring was part of the way liable for the development of stone tors.

Deferred Junction

A kind of waterway conversion in which the intersection of a feeder with a standard on a flood plain is deferred by the presence of a levee. The feeder may in this way be compelled to run corresponding to the standard for a significant distance downstream. An outrageous illustration of a conceded intersection is the Yazoo, which streams close by the Mississippi for 280 km before really going along with it.

Deficiency Disease

An infection caused where some component indispensable to human wellbeing and endurance is absent. Most regularly, the insufficiency happens when the diet does not have a particular mineral, protein or nutrient. For instance, rickets (a bone delicate quality happening in kids) is brought about by an inadequacy of vitamin d.

Deflation

The expulsion of fine material by wind activity, especially in deserts and along coasts. It is normal viewed as a type of wind erosion,' however truth be told it includes just the vehicle of relaxed particles, primarily sand. Emptying is normally a limited scale and inadequate interaction, however when usable over a huge region its complete impact can be critical. Resultant landforms are generally little hollows and blowouts.

Deglaciation

The decrease in size of a glacial mass or ice sheet, coming about because of a negative mass balance and prompting the openness of the already ice-covered surface. Deglaciation includes both retreat of the ice edges (as ablation surpasses positive headway) and down wasting of the ice surface. Since it delivers enormous amounts of meltwater, different landforms related with lakes and fluvial action are shaped, along with recessional moraines.

De-Industrialization

The decay (relative or total) in the significance of assembling experienced by most MEDC (More Economic Developed Country)since the mid-1960s, as estimated by its commitment to GDP and its level of the all-out workforce. It is clarified by declining usefulness (maybe because of old plant, prohibitive work rehearses, high charges, a solid money, and so on) and solid unfamiliar rivalry, especially from nations where work is less expensive and unionized.

The USA and UK have been among those nations generally influenced to date, and their spaces of weighty (iron and steel, shipbuilding) and conventional (materials) industry have been hardest hit. The decay of the assembling area is stressing to certain legislatures, since they accept that the tertiary sectors and quaternary sectors are unequipped for raising a similar degree of abroad income. A lot of the assembling lost by MEDC has been moved in LEDC with their attractions of less expensive land and work, and less rigid controls on contamination.

Demographic Regulation Theory

This expresses that each general public will in general keep its richness and mortality in a kind of equilibrium. At the point when the death rate is high, there is little requirement for the guideline of the birth rate. Yet, when mortality is decreased to low levels, then, at that point the steadiness of high ripeness (and in this manner of high paces of natural increase) will create a size of populace development that may well surpass the carrying capacity of the land. This, thus, will be seen as taking steps to bring down broad well-being. In these conditions, the guideline of richness gets acknowledged as something waiting be attempted for the aggregate great. This hypothesis is significant in clarifying the demographic transition.

Dendritic Drainage

A typical sort of waste example, related with spaces of flat or delicately plunging layers. Dendritic seepage contains a huge number of little tree-like branch streams that join together, for the most part at an intense point, to give bigger streams and, at last, one significant trunk stream. Generally, the example takes after that of the parts of an enormous tree. The real closeness of the example will shift enormously, contingent upon rock penetrability, and the sum and nature of precipitation. It is likewise prone to fluctuate through time.

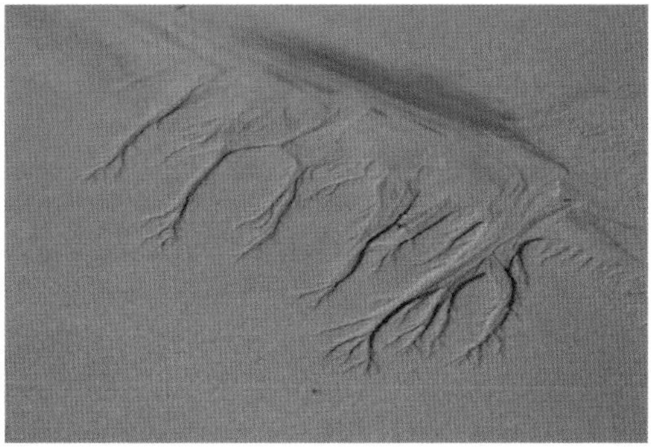

Figure 16. Pattern of dendritic drainage.

Source: Image by Wikimedia.

Dendrochronology

The study of recreating past climatic changes, utilizing the proof of tree rings, every one of which reflects (in its width) the measure of development as controlled by the temperature and precipitation of the year where it was framed. Tests from living trees (acquired from centers of wood separated from the tree trunk) and long-dead trees can be collected to give a climatic record over a time of a few thousand years. Singular tree types of incredible life span, for example, the bristlecone pine of the Sierra Nevada in California, can give a record of up to at least 4000 years.

Denudation

A term that refers to the load of cycles of weathering. Transport and erosion that are liable for the bringing down and forming of the actual scene. The term denudation sequence alludes to the investigation of the advancement of the scene to its current structure throughout extensive stretches of time, as shown by past 'denudational occasions' like the arrangement of erosion surfaces, marine stages, river terraces, and so forth, and related stores.

Dependence

The condition in which a country, locale or gathering of individuals is simply ready to endure and advance by dependence on help gave by another. This help can take different structures, including exchange, aid, political assurance and protection. Reliance at a public level, especially in the third world, has been the result of colonialism. Conventional social orders were initially sensibly independent – it was the appearance of European settlers that changed over them into subordinate nations.

Dependent Variable

A scope of qualities which, in correlation and regression analysis, are thought to be connected here and there to another independent variable. In the event that information concerning levels of SO2 in the atmosphere are acquired from various locales and an endeavor is made to relate these figures to information about the consuming of petroleum products, then, at that point those utilization esteems would comprise the autonomous variable and the SO_2 information the reliant worth (e.g., thought to be identified with fuel utilization).

In the development of a graph to show the connection between the two factors, the ordinate or Y pivot would be utilized to plot the so_2 information and the abscissa or X hub the fuel utilization figures. The exact connection between the two factors may likewise be set up by measurable strategies (e.g., relationship coefficient, rank correlation).

Depopulation

A decrease in the quantity of individuals living in a space, all the more normally achieved by net migration deficit as opposed to by unreasonably high mortality. Such misfortunes from country regions might be alluded to as rustic termination.

Deposition

Most generally, the setting down of sediments delivered by the weathering and erosion of land-masses. Likewise, testimony can include compound precipitation in seas and lakes (shaping evaporite stores or limestone), and plant development and rot.

Statement can occur in an assortment of conditions (fluvial, marine, aeolian, and so forth) and is generally constrained by the size of particles being conveyed and the speed of transport. Bigger particles are stored at higher speeds in moderately high-energy conditions (e.g., on a tempest sea shore), while better particles are stored at lower speeds in low-energy conditions (e.g., in a protected tidal pond).

Depressed Region

A locale that, in monetary terms, is performing less well than the public normal, and regularly progressively so. A locale described by monetary pain and decrease (thus the frequently utilized elective assignment declining district), as shown by indications, for example, moderately significant degrees of joblessness; outpouring of capital; a stamped withdrawal of a fundamental industry on which the region has since quite a while ago depended (e.g., the shipbuilding business in new England); conclusion of firms, frequently connected to the essential business; migration of work.

When begun, a kind of opposite multiplier effect or descending twisting (it is made, inciting steady disintegration in the area's infrastructure and public administrations. This, thus, will make further joblessness, empower more out-relocation and lessen speculation certainty. In many MEDC, such locales get different types of government help. Regularly this is essentially for the political catalyst of apparently being effectively check a decay that, by and by, is amazingly hard to invert. In the core-periphery model, discouraged locales would be viewed as descending change districts.

Deprivation

An idea applying to the two individuals and regions, in light of the thought of burden comparative with others and different regions. Hardship relates especially to the three domains of lodging, work and administrations. For instance, there are a few areas of society denied of the chance: to live in good lodging; to acquire a satisfactory compensation; to have legitimate admittance to different administrations. Accordingly, hardship has various features, and these will in general be interrelated, for instance, absence of a task or of a satisfactory pay forestalls admittance to legitimate lodging; subsequently the often-utilized term, numerous hardships, similarly, hardship is a spatial marvel in that there are entire regions having a tendency to pass up a major opportunity – for example, discouraged regions and downtown.

Desert

A bone-dry district or area that encounters extremely low yearly precipitation and, accordingly, has almost no vegetation. The majority of the world's abandons lie between 15–30° N and S of the equator, relating with the sub-tropical belt of high pressing factor. Here the air is by and large sinking, subsequently the absence of convective action prompting cloud development and the absence of downpour. With a shortfall of cloud, diurnal temperature ranges are extremely high; during the day temperatures will frequently surpass 30°C, though around evening time ices may happen, particularly in the colder time of year. While the term 'desert' is regularly connected with hot deserts, it additionally applies to districts of low precipitation in polar regions, for example, huge pieces of antarctica.

Figure 17. Sahara desert.

Source: Image by Wikimedia.

Desert Pavement

A stony desert surface, framed where sand particles have been taken out by deflation, leaving a firmly stuffed layer of rocks (slack rock), which are themselves regularly faceted by wind abrasion. The stones might be established together by minerals attracted upwards to the surface in arrangement and encouraged because of dissipation.

Desertification

The spread of desert conditions into previous spaces of semi-parched hedge, steppe field and even forest. For instance, in ongoing many years, the southern edges of the Sahara have been progressing at a disturbing rate, especially during and since the incomparable Sahel dry spells of the mid-1970s. While desertification can to some extent be connected to 'regular' climatic changes, it is progressively acknowledged that the exercises of individuals (for instance, overgrazing of homegrown creatures, overcultivation and the enormous scope annihilation of forest for kindling) is a significant contributory factor, or perhaps even the fundamental driver of desertification.

At the point when land becomes decertified, it is not, at this point ready to help its kin. Food deficiencies result and individuals are compelled to relocate to different districts, relinquishing their property and their jobs. At the point when the inconsistent downpours do show up, frequently as deluges, genuine soil erosion can result as the land has been deprived of its defensive vegetation layer. Stream channels become silted and the flood danger increments. Desertification is presently viewed as one of the best and most earnest issues facing numerous tropical nations.

Desire Line

In transport examines, a straight line drawn on a guide joining two focuses between which there is a longing or motivation to travel, however not really the genuine course to be followed. On a longing queue chart, one line is drawn for each different development, touching to the following, so the general thickness of the lines gives a visual sign of the absolute number of wanted excursions.

E

Earth Flow

A sort of mass movement in which free incline materials become immersed with water and stream at moderate to exceptionally fast rates. It might happen at the toe of a landslide, attributable to the focus there of soil water. Where the stream influences primarily clay-sized particles with a high dampness content, an earth stream turns into a mud stream.

Earth Movement

The heat from radioactive processes within the earth's interior leads to the movement of its plates the direction of which varies – sometimes toward and sometimes away from each other. This movement is called plate motion, or tectonic shift. In other words, earth movement refers to a force inside its outside layer because of compressive or tensional powers. The movement of these plates is shown by earthquakes and slight earth quakes.

Earth Pillar

Earth pillar refers to a forcefully pointed apex, created in a mass of broken stone or garbage exposed to extreme erosion by rain-wash and little streams. Earth columns are especially striking in certain pieces of the alps, like the pennine alps and the Tyrol, where old lateral moraine stores containing boulders have been profoundly gullied following icy mass downturn. As erosion continues, the biggest rocks are uncovered and afterward go about as 'umbrellas,' protecting the better material underneath from downpour sprinkle disintegration. The stones are hence left progressively upstanding as 'covers' to columns as the encompassing material is washed away.

Earth Summit

The name given to the gathering held in Rio de Janeiro in 1992 by global pioneers and protectionists who were worried about major natural issues like global warming, the ozone layer, deforestation and the quick exhaustion of non-inexhaustible resources. In the event that these issues are to be survived, it is broadly accepted that administrations should seek for strategies of sustainable development.

Earthquake

An unexpected and brief time of exceptional ground shaking. Seismic tremors are typically connected with volcanic action or development along a fault. The starting place of a seismic tremor is known as the core interest. The point on the ground surface promptly over the center is known as the focal point. At the point when a tremor happens, stun waves (seismic waves) travel outwards like waves in a lake.

There are three kinds of wave:

i. essential or p waves, which are the quickest;
ii. auxiliary or s waves;
iii. long or L waves, which notwithstanding being the slowest are liable for causing the best obliteration since they travel through the hull as it were.

Quakes are recorded by seismographs and the time-stretch between the appearance of p and s waves at a seismic station can permit estimation of distance to the tremor center and season of event.

Figure 18. Destructions caused in Haiti earthquake.

Source: Image by flickr.com.

The most impressive quakes are related with significant separation points – for instance, the San Andrea's 'megashear,' which led to the incomparable San Francisco tremor of 1906 and the extreme Turkish quake on the Anatolian fault in 1999. The extent of tremors (a term for the all-out energy delivered by a quake) is estimated by the open-finished Richter scale, where each point on the scale addresses a x30 expansion in the measure of energy discharge. The impacts of a quake can be estimated utilizing the modified Mercalli scale, which has 12 degrees showed by roman numeral.

Ecological Balance

A 'consistent state,' accomplished by an ecosystem wherein inputs have come to approach yields; this is likewise alluded to as homeostasis. An illustration of biological equilibrium is climatic climax vegetation, in which the 'regular' vegetation has gotten consummately acclimated to the predominant climatic conditions. Nonetheless, it is currently generally acknowledged that natural equilibrium is to a greater extent a hypothetical idea, since controlling elements, for example, environment and soil are continually going through change.

Ecology

The logical investigation of the common connections of plants and creatures to their environment (a differentiation is some of the time made, for comfort, between plant biology and creature nature). A scientist is concerned especially with the natural cycles at work inside the ecosystem, remembering the ways for which creatures acquire energy and matter both from the actual climate and one another, and thus discharge that energy and matter back to the climate.

Economic Growth

A term alluding to the outer conditions where firms work and which incorporate such components as financing costs, market lightness, joblessness levels, charges, government endowments, and so on essentially, the financial climate covers conditions over which the firm has no genuine control, however to which it should react. The elective terms monetary environment and business environment are broadly utilized.

An essential part of development, including rising degrees of material creation and utilization, and as often as possible changes in the sectoral equilibrium of the economy (e.g., the secondary sector and the tertiary sector acquire in significance). The advancement of financial development might be observed by a scope of measures, for example, pay, GDP and GNP per capita, mechanical creation, levels of administration arrangement, energy supply and utilization, exchange, capital streams, and so on thusly, these actions demonstrate that financial development is, in itself, a complex interaction of progress, the exact course and character of which will fluctuate from one spot to another, contingent upon the nature and nature of resources (physical and human) and on promising circumstances.

Economic Indicators

An economic indicator is a statistic or an information presented in numbers about an economic activity. Economic indicators are designed in a way that they allow analysis of economic performance and also help in predicting future performance. Insights that are vulnerable to changes in the condition of industry, exchange and business inside the public or territorial economy. In the UK, monetary markers incorporate the insights of joblessness and unfilled occupation opportunities, gold stores, bank propels, yield of steel and engine vehicles, balance of trade.

Economic Overhead Capital

Economic overhead capital relates to infrastructure such as roads, power transmission systems, telecommunications, etc. A typical characteristic or feature of economic infrastructure is its high fixed cost and relatively low variable costs of operation. In other words, it simply refers to capital resources that are put into infrastructure. It also relates to economic development.

Economic Planning

The distinguishing proof of future monetary necessities, and the association and arrangement of scant resources in the most effective manner to fulfill those requirements. In this sense, monetary arranging can be attempted by the individual, firm, nearby position, local and public governments. In entrepreneur social orders, financial arranging will in general appear as shifting levels of government intervention, especially as for provincial monetary turn of events (e.g., rectifying regional imbalance), and controlling financing costs and assessments.

Economic Rent

The contrast between the absolute income got from the offer of a ware and the complete expenses of creation and transport. As utilized in agricultural geography, financial lease is the excess return or benefit coming about because of utilizing land for one sort of creation as opposed to another. The income got is controlled by the market cost for that product and this, thus, is a reaction to the market interest circumstance. Creation costs are thought to be fixed (e.g., they don't fluctuate from one spot to another), while transport costs increment with distance from the market.

The more noteworthy the vehicle costs, the more modest the contrast among income and absolute expenses, and accordingly the more modest the monetary lease. In this manner, financial lease diminishes with distance from the market, however the monetary lease slope will fluctuate as indicated by the sort of horticultural creation.

Economies of Scale

Economies of scale refer to the cost advantage experienced by a firm or an organization or by a country when it increases its level of output for a particular commodity. The advantage arises due to the inverse relationship between per-unit fixed cost and the quantity produced. The greater the quantity of output produced, the lower the per-unit fixed cost.

Ecosystem

A natural local area of plants and creatures saw inside its actual climate or habitat, for example, a freshwater lake, a blended forest, or a support. An environment can be depicted as a 'complex of collaborating marvels,' inside which there are many convoluted and frequently unpretentious connections (among environment and vegetation, vegetation and soils, creatures and vegetation, etc.). It is feasible for an environment to get steady, with the different parts accomplishing a state of equilibrium or harmony. Be that as it may, it is likewise feasible for biological systems to turn out to be genuinely upset, or shaky, because of normal disasters – for instance, a significant volcanic eruption or impedance by individuals.

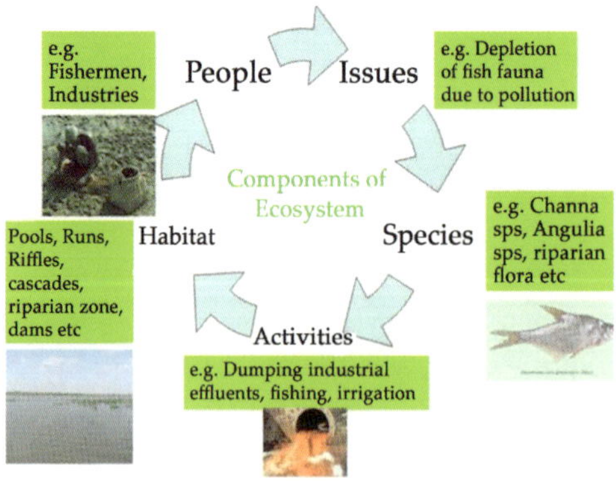

Figure 19. Components of ecosystem.

Source: Image by Wikimedia.

Ecotourism

Ecotourism can be understood to be a form of tourism that involves responsible travel to natural areas, with the key objective of conserving the environment, and improving the well-being of the local people.

In other words, it is a type of tourism that looks to limit the otherwise natural/harmful impacts of tourism pressure and the development of infrastructure to cater to the rise in tourism. It's anything but a response to the enormous ecological, social and social harm that has been brought about by the travel industry. A significant part of ecotourism is that it ought to be claimed and run by nearby individuals and the benefits held in the neighborhood than 'spilling' to an unfamiliar organization or place.

Edaphic

A term portraying the soil conditions that impact the development of plants and different organic entities. These incorporate soil surface and causticity, the presence or nonappearance of essential minerals (counting minor components) and the soil moisture content. Plant geographers perceive that, while significant vegetation districts are climatically resolved, inside those areas' neighborhood varieties in vegetation (plant communities) regularly mirror the impact of edaphic factors.

Edge

A term utilized in graph theory to portray the connection between two vertices (nodes) of a topological map. The edges (now and then likewise alluded to as connections) of a vehicle network would be the courses connecting settlements.

Edge City

The result of metropolitan cycles (mostly decentralization) prompting portions of the suburbs getting greater city-like through the agglomeration of workplaces, industrial facilities and huge shopping buildings at supported, available areas. They are highlights of post-rural America, yet are starting to show up in Europe.

Electricity

Electricity is defined as the flow of electrical power or charge. It is not a primary, but a secondary energy source as it is produced or achieved by the conversion of other sources of energy, like coal, natural gas, oil, nuclear power and other natural sources. These are called primary sources.

The energy sources that are used to generate electricity can be either renewable or non-renewable, but electricity itself is neither renewable or non-renewable, it is basically a part of nature and one of the most prominently used forms of energy. Historically, many cities and towns across the globe were built alongside waterfalls (which is a primary source of mechanical energy) that turned water wheels to perform work of generating electricity (also known as hydro power generation).

Eluviation

In everyday utilization, the washing down of soil constituents through a soil profile by the way toward leaching (including the dissolving of supplements) and mechanical downwash (including solids like humus or clays). Eluviation is advanced by high precipitation and acidic vegetation, and is related with the arrangement of podzols. Materials might be re-saved to shape an illuviated lower skyline.

Emanant Coast

A coastline that has risen, or is as yet rising, comparative with ocean level. Rise can be because of vertical elevate of the land – for instance, because of isostatic recuperation following the dissolving of a huge ice sheet, or a genuine fall in sea level, for example, at the beginning of a cold development. Commonplace highlights of a developing coast are deserted precipice lines, raised beaches, and broad waterfront fields addressing the previous ocean bottom.

Endangered Species

An endangered species is a species that has high chances or probability of becoming extinct in the near future, either worldwide or in a particular geographic or political jurisdiction. Endangered species may be at risk due to factors such as habitat loss, poaching and invasive species.

An increasing number of plant and creature species are at risk of eradication as a result of human intervention through the exploitation of natural habitats. Misuse of natural resources, and so on in excess of 1000 creature species are recorded in the red book delivered by the international union for conservation of nature and nature reserves, while more than 25,000 types of plant are on the endangered list. This significant issue, with its subsequent loss of biodiversity, requires a lot more attention. It likewise needs globally concurred conservation programs and a major reappraisal of the connection between the human species and the normal world.

Energy Crises

The purported oil emergencies date from 1973 and 1979–80 when the Libyan government, trailed by the remainder of OPEC, tested the world oil-creation and oil pricing strategies, which up to that point had been set somewhere around the worldwide oil organizations. In many middle eastern nations, the resources and tasks of those organizations were seized and nationalized and, despite the fact that oil was never hard to come by, those nations keenly designed the circumstance so as to make the cost of raw petroleum ascend in a dynamite way.

Undoubtedly, somewhere between 1971 and 1981 the free cost of oil expanded by almost multiple times, and oil turned into a powerful weapon in the legislative issues of global connections. The impacts of the emergency were most distinctly felt by those nations subject to imported oil supplies, especially those third world nations with few resources to offer in return. It additionally seriously influenced Japan, which determined more than 75% of its all-out energy necessity from middle eastern oil.

It empowered the global oil organizations to attempt investigation for oil in those pieces of the world considered to be politically steady (e.g., liberated from the danger of seizure), however that are earth troublesome (e.g., the frozen north, the North Sea). It stirred interest in the more conventional fills, like coal, just as inciting R&D into conceivable alternative energy dependent on inexhaustible assets (e.g., wind, wave, flowing and sun-based force). Moreover, it's anything but various nations to set out on mechanical rebuilding.

Environmental Lapse-Rate

The genuine abatement of temperature with elevation in the world's atmosphere over a specific spot at a particular time. The natural slip by rate (ELR) has a mean worth of 0.6°C per 100 m. It changes with the section of various air masses. For instance, polar sea air warmed from underneath in its section southwards will in general have a 'precarious' ELR over Great Britain, while tropical sea air cooled from beneath as it moves northwards has a 'delicate' ELR. ELR additionally fluctuates diurnally. During the daytime, as the ground is warmed by sunlight-based radiation, it steepens, in this way improving the probability of barometrical shakiness. Around evening time, it is decreased with the cooling of the ground and the overlying air, accordingly preferring dependability. While temperature by and large reductions with height, there might be events where it stays steady (isothermal slip by rate) or really increments to frame a temperature reversal.

Environmental Pollution

The aggravation of the common habitat coming about straightforwardly and by implication from human exercises. The causes or wellsprings of contamination are many, going from the air emanations related with the copying of non-renewable energy sources to the use of fertilizers and pesticides, from the release of mechanical and homegrown profluent into streams and other water bodies to the age of terrible commotion levels.

The results of contamination are similarly different, going from the making of acid rain to rising degrees of mercury in the ocean, from the breaking or interruption of food chains to the hazarding of human wellbeing. Lately, there has been a developing public mindfulness in numerous nations of the genuine size of ecological contamination, and projects of natural insurance are getting ordinary in spite of the fact that they are not in every case simple to police.

Epicenter

The point on the outside of the earth lying promptly over the focal point of an earthquake. The last typically happens at a profundity of 0–50 km; be that as it may, 'profound center's tremors, with a profundity of beginning more noteworthy than 250 km, have been recognized. The world appropriation of focal points shows a stamped fixation around the edges of the Pacific Ocean, along the middle line of the Atlantic, and through the Mediterranean bowl into turkey, Iran and past. As such, along plate edges. By and large, the best annihilation related with a seismic tremor happens near the focal point. There are, be that as it may, exemptions. In the 1985 quake that harmed enormous pieces of Mexico City and killed 7000 individuals, the focal point was many kilometers away.

EPZ (Export-Processing Zone)

A generally little, obviously characterized region inside a country in which the point is to draw in export-oriented businesses by extraordinary concessions, for example, exception from charges and obligations, especially as to the import of crude materials and parts utilized in the creation of products for send out; arrangement of the vital actual foundation; no limitations – on unfamiliar possession. There are currently more than 200 EPZ in activity, the greater part of them set up after 1971. Eminent groups happen in Central America and the Caribbean; Singapore is basically one enormous fare preparing zone, as was Hong Kong before it was gotten back to China.

F

Fabric Effect

The manner by which components of the townscape acquired from the past influence present-day land use. For instance, old structures may not loan themselves to new uses, and accordingly their protection will oblige land-use change. Essentially, the conservation of old and restricted road frameworks may obstruct the development of traffic.

Factor Analysis

A factual system that actions the evident interrelationships (covariance) of selected variables. In topographical exploration such factors are generally recorded either at a progression of various areas (e.g., researching the qualities of a bunch of soil tests taken at better places) or over an organization of sub-regions (e.g., examining the connection between population density, socio-economic status and lodging conditions in the wards of a town).

Factors of Production

The fundamental components of the creation interaction; those things that are important before creation can start. They are by and large viewed as being triple: labor, land and capital. Some would perceive enterprise as a fourth factor. Work necessities differ as indicated by the sort of creation, while land is central in an assortment of ways, for example, as a wellspring of raw materials, giving mechanical locales, and so forth

Factory Farming

An arrangement of domesticated animals cultivating in which creatures are raised seriously under cover and in states of limited versatility. Poultry, pigs and calves are regularly raised along these lines. Plant cultivating is pointed toward expanding yield, diminishing expenses and delivering a normalized food item, regularly sold through store chains.

Famine

A shortage of food, prompting malnutrition and starvation, incited fundamentally by some disappointment in the food creation measure, for example, by flooding (as in Bangladesh); by determined drought (as in the

Sahel); by the assaults of infections and vermin (like grasshoppers); by political agitation (as in Somalia). Overpopulated regions will in general be particularly defenseless against such disappointments. Starvation may likewise be accelerated by deficiently exhaustive aid programs, which might be viable in lessening levels of mortality, yet are discovered needing with regards to creating the additional food needed by the subsequent expansion in populace.

Figure 20. Image showing Ethiopian Famine.

Source: Image by flickr.com.

Fault

A break in a stone, initiated by either tensional or compressive powers, in which there is relocation along a separation point. A deficiency can be recognized from a joint in that, with the last mentioned, no relocation happens. The side of the deficiency that has encountered a general fall is known as the downthrow side; the opposite side is named the upthrow side. Among the principle kinds of shortcoming are: ordinary issues, created by strain and including bringing down of the stones as an afterthought towards which the issue plane is plunging; switched issues, coming about because of pressure, which raises the stones as an afterthought towards which the issue plane is plunging; tear issues (also called change issues), in which the development of the stones on one or the other side of a close upward flaw plane is essentially even and corresponding to the deficiency. Blaming is straightforwardly answerable for the creation of certain landforms, and furthermore frames lines of shortcoming that can, for instance, be misused by erosional cycles to shape valleys.

Fetch

The distance of vast ocean over which wind hits to create waves. Bring is a central point controlling the tallness and energy of waves, and consequently their capacity to alter seaside landforms by erosion, transport and deposition. The bearing of most extreme bring, which assists with deciding the methodology of the dominant waves, is especially significant inside encased or incompletely encased oceans.

Firm

A unit of the board working under a business trademark and coordinated to take part in financial action (be it mining, assembling, or selling labor and products). A firm might be a sole ownership, a private or restricted risk organization or a state-claimed venture.

Firn

In a real sense signifying 'last year's snow,' it is snow that has lain for a year without softening. Firn addresses the equilibrium of the colder time of year snowfall on the accumulation zone of a glacial mass after the expulsion of the highest layers by summer ablation. The excess snow then, at that point gets covered and compacted by the following winter's snowfall. In this manner the icy mass is developed by progressive augmentations of firn, which go through slow change, including recrystallization and the prohibition of contained air bubbles. Accordingly, the thickness of the firn expands, the ice precious stones become bigger, and in the end, it is changed into genuine ice sheet ice. The pace of progress fluctuates with nearby conditions (e.g., the thickness of winter snowfall, the measure of defrosting and refreezing throughout the mid-year season, winning temperatures, and so forth) and is a lot of faster in mild than polar icy masses. The firn line is the lowermost furthest reaches of firn on a glacial mass and addresses the limit between the gathering zone and the ablation zone.

Fissure Eruption

A direct volcanic eruption in which liquid lava arises in huge amounts and with minimal unstable action, along broad lines of crustal shortcoming – faults, breaks or significant joint lines. Throughout some stretch of time progressive magma streams develop, now and then to thicknesses of many meters, and 'immerse' the prior relief highlights, like valleys and interfluves. The last are supplanted by broad plateaus (typically involving basalt).

Figure 21. Fissure eruptions.

Source: Image by pxhere.com.

Fixed Costs

i. Costs that don't differ with the volume of creation, for example, costs not influenced by economies of scale (e.g., overhead expenses like lease, interest in plant and apparatus).
ii. Costs that are consistent in space and in this way have no effect on comparative advantage (e.g., monetary capital or the expenses of work and materials where broadly concurred rates apply).

Flash Flood

A fleeting flood, portrayed by: an extremely fast beginning (or 'steeply rising hydrograph'); a short time of pinnacle stream; a moderately sluggish decrease in waterway discharge ('downturn period'). Streak floods are especially connected with deserts, where downpour is regularly as short (rare) extreme showers, the ground surface is 'prepared' and impermeable and there is little vegetation to block the progression of water.

They are additionally normal for rugged conditions where the lofty inclinations, perhaps joined with deforestation, send water quickly. In December 1999 the shoreline of Venezuela endured extreme glimmer flooding following heavy precipitation; nearly 50,000 individuals were thought to have been killed by the floods and the related landslides. Streak floods play out a significant job in shipping sediments (in suspension and as bed load) over moderately brief distances; in any case, as release is decreased and silt fixation builds there is a fast change from transport to deposition a downstream way.

Flood

A flood is an overflow of water on otherwise a dry ground. Floods are most often caused by an overflowing river, a dam break, snowmelt, or heavy rainfall, and sometimes, quite less frequently, by tsunamis and storm surge. Floods can also be defined as a time of high discharge of a stream, coming about because of conditions, for example, substantial precipitation; serious liquefying of snow and ice; the penetrating of normal boundaries, (e.g., ice dams); the breakdown of counterfeit barrages. Stream floods have been characterized as 'occasions' of such greatness that the channels can't oblige the pinnacle release.

At the end of the day, a flood is a stream in overabundance of the channel limit, and results in immersion of low-lying level land contiguous the channel. Floods may happen occasionally (as on the Blue Nile) or at more unpredictable spans relying upon the event of individual extreme focus rainstorms, for example, that creating the cataclysmic flood at Lynmouth, DeVon, England, on 18 August 1952.

One significant angle is the expectation of stream floods, to help safe designing development of scaffolds, waterway dikes, and so forth and to stay away from the siting of houses in regions subject to genuine flood perils. From an investigation of past records, an endeavor is made to decide the repeat time frame of specific measurements. In basic terms, the biggest flood that happened during the previous 50-year time frame is probably going to be coordinated by a relating flood during the following 50 years.

In all actuality expectation is probably going to be undeniably more confounded for an assortment of reasons. The accessible time of study may not accept the genuinely remarkable flood (the '1000-year flood'); however, it is possible that such a flood could happen whenever sooner rather than later. Another factor is the change of waterway catchments by individuals' activities in deforestation, agriculture, land waste, urbanization, and so forth, which may impressively adjust the 'likelihood' of surges of a specific size.

The term flood is likewise utilized from a more extensive perspective to allude to the immersion of land by other than stream water – for instance, because of high lake levels coming about because of astoundingly high precipitation, or unusually high ocean levels, as during the 'storm floods' in England in 1953.

Flood Plain

That piece of a valley floor over which a stream spreads during occasional or transient floods. During such occasions the speed of stream is not exactly that inside the waterway channel, and in the moderately slack water over the plain suspended sediment gradually settles out. Accordingly, the flood plain is gradually developed by additions of sand, silt and clay. The flood plain is additionally altered by changes of the waterway direction (in the turn of events and relocation of meanders), and flood plain stores likewise include material from broadened point-bars. By definition, flood fields are spaces of delicate relief, giving an impression in the field of practically amazing levelness. In any case, exhaustively they are differentiated by muddy sorrows checking deserted channels and oxbows, and along the stream edges confined deposition may bring about levees.

Food Chain

A progression of creatures with interrelated taking care of propensities, every living being filling in as nourishment for the following in the chain. The chain typically starts with creatures that produce vegetal matter, via photosynthesis including daylight (sun-oriented energy). Consequently grasses, spices and bushes give food to nibbling and perusing creatures (herbivores like gazelle, zebra and wildebeest), which are thus gone after via carnivores (panther, lion and hyena). Once more, plants and seeds are devoured by little and medium sized birds (sparrows, pigeons), which are gone after by flying predators, for example, sparrow birds of prey and peregrine hawks. In actuality, there is a persistent, yet consistently diminishing, move of energy through the natural pecking order.

Figure 22. Complete circle of food chain.

Source: Image by Wikimedia.

Food Production Chain

The arrangement of connected stages that beginnings with the rural contributions to cultivate creation, and finishes with food appropriation and utilization. It is the focal string going through the overall food supply framework.

Footloose Industries

Businesses that don't appear either to be attached to any exceptional sort of area or to have any superseding locational prerequisites. Firms engaged with a similar free industry (e.g., light designing) are regularly to be found in various sorts of area. It is guaranteed that the advancement of network frameworks for the circulation of energy and force has would in general make industry all in all fairly freer.

Foreign (Overseas) Investment

Embraced by firms as they develop, to expand their business advantages and benefits by expanding creation, broadening markets or differentiating.

There are three primary structures: (i) setting up branch plants and auxiliary organizations (as by the Japanese organization Sony in the UK); (ii) going into joint endeavors with unfamiliar organizations and Honda); (iii) procuring value (shares) in unfamiliar organizations. Internal venture alludes to the above as seen from the 'getting' country. Unfamiliar speculation is a lot of part of globalization and the 'matter' of TNCS.

Forestry Commission

Set up in 1919 to redress the enormous scope deforestation of Britain previously and during the first world war and to develop the country's load of lumber for what's to come. A significant part of the 1.2 million ha of land held by the commission is offered over to coniferous plantations. Albeit effective in renewing wood stocks, throughout the long term its work has been reprimanded on various tallies. For instance, the discount planting of speedy developing, however 'outlandish' conifers has implied adjusting the customary appearance of the rural scene (an especially combative issue in the English lake district).

The commission is additionally dynamic in fields other than financial ranger service. For instance, it is liable for the administration of the new forest, where its time and assets are likewise coordinated towards the consideration of deciduous forests, the support of touching for enormous quantities of domesticated animals and the arrangement of sporting framework (vehicle leaves, excursion regions, campgrounds, and so forth) the as of late settled forest enterprise is a business exchanging arm of the forestry commission that deals with the ranger service bequests on a multi-use premise.

Freight Rates

The expense of shipping an item over a given distance. Cargo rates shift as indicated by: the method of transport; the sort or amount of the ware to be moved. The consideration of terminal costs in the estimation of cargo rates implies that normal vehicle costs per kilometer reduction or show a tightening as the effect of distance on absolute vehicle costs linear relationship, complete expense increment relatively with distance curvilinear relationship, all out costs not expanding relatively with distance (e.g., tightening) stepped relationship, costs don't increment relatively with distance and they expansion in stages no increment of expenses with distance length of take builds. It ought to be recalled, in any case, that there are some cargo rate structures that work regardless of distance, with respect to model postage. There are likewise cases where ventured duties apply, with a uniform rate being received inside a given zone.

Frequency Distribution

The recurrence appropriation of a variable alludes to the quantity of events of various qualities. Such recurrence information should be estimated either on a consistent scale (e.g., stream speed) or gathered into classes (e.g., the quantity of settlements in a given nation falling into every one of a predefined arrangement of size classes). Plotting such information on a graph delivers either a recurrence bend, on account of nonstop information, or a histogram on account of gathered information. There are numerous kinds of recurrence bend; they incorporate ordinary, slanted and bimodal. These terms portray the general 'state' of the dissemination.

Frontal Depression

A space of low climatic pressing factor, created along the polar front and described by warm and cold fronts, and a warm sector. Dejection's structure most promptly over the seas in mid-scopes and track eastwards, carrying mists and downpour toward the western edges of landmasses. At a front, warm clammy air is compelled to rise. As it does as such, it cools, prompting condensation of water fume, cloud development and precipitation. At the warm front (of generally delicate angle), inspire is less quick however stretches out over a wide region; the resultant precipitation is consistent and may proceed for a few hours. At the virus front (of moderately steep inclination), inspire is faster and influences a lot more modest region; the resultant precipitation is more serious, brief and ordinarily includes substantial showers.

Frost Heave

The vertical development of rock or soil particles because of the pressing factors created by the arrangement of ice isolations in the ground. Regularly there is likewise a level of parallel dislodging (named ice push) with the end goal that the term ice extension might be the most fitting to utilize. Ice hurl takes a few structures; for instance, the development of ice gems (particularly needle ice) will elevate soil particles; and bigger ice bodies may raise enormous stones. One of the chief aftereffects of ice hurl is the upward arranging of flotsam and jetsam, with the coarse particles being raised to the surface above hidden fines.

Frost Hollow

A low-lying discouragement or valley into which cold air channels around evening time, especially under the stable climatic conditions related with anticyclones. On the nearby inclines, nighttime radiation of warmth makes the ground become cold. The overlying air is thusly cooled and, being moderately thick, it streams downhill to amass in the downturn or valley. Ground temperatures on the inclines may stay above freezing, yet in the empty they may fall well underneath it, so the ground surface and vegetation become white with ice. For sure, temperatures inside an ice empty may tumble to an exceptionally low level, especially in a valley where the progression of air along the floor is obstructed by a street or railroad bank. Organic product cultivators specifically look to keep away from ice hollows, since the ice will cause genuine harm should it happen during bloom time.

G

Gaia Hypothesis

The Gaia hypothesis, or the Gaia theory, the Gaia paradigm, or the Gaia principle, is of the view that living organisms interact with their inorganic surroundings on Earth to form a synergistic and self-regulating, complex system that helps to maintain and multiply the conditions for life on the planet. In other words, it's an idea where the earth is seen as a mind-boggling living being equipped for directing and coordinating its numerous segment parts and cycles. It is especially valuable in understanding the unpredictable changes that keep up the planet's drawn out climatic balance.

Gap Town

A town arranged in, or close, or directing, a hole in an edge – for example, Guildford and Dorking in the UK's north downs. Hole towns regularly owe their starting point not exclusively to their capacity to order courses through the hole along the valley floor, yet in addition to their deserving situation with admiration to courses crossing the hole from key position on one or the other side of it. Antiquated courses regularly ran cross over to the hole, not through it.

Garden City

A name given by Howard (1898) to depict a cautiously and completely arranged town, intended to keep up something of an open rural person, with generally low lodging densities, and with specific consideration paid to the area of industry, administrations and amenities. The aim was to consolidate the benefit of town daily routine with the attractions of experiencing in a solid provincial climate. Howard delivered a model of a nursery city with the pronounced point of making 'urban areas in nurseries, and nurseries in urban areas.' The best city size was considered to be around 30,000 occupants.

Howard and his adherents (known as the garden city movement) endeavored to exhibit the achievability of the model by embraced two nursery city adventures in Hertfordshire: one at letch worth (began 1903) and the other at Walwyn (began 1920). The use of a similar plan thoughts to a rural setting was shown by the development of Hampstead Garden suburb (began 1907). These examinations spearheaded thoughts of metropolitan plan later applied in the British new towns and lodging homes of the after-war time.

Figure 23. The garden city in Singapore.

Source: Image by flickr.com.

Garigue (Garrigue)

A savage type of scour forest related with slim, poor and dry soils, and particularly normal for limestone regions around the Mediterranean. The vegetation is low-developing, sketchy and contains numerous fragrant plants (savvy, lavender and thyme). In the driest regions, even obvious desert plants (like thorny pear and aloe) are found.

Gatekeepers

In human geography the term alludes to proficient individuals (domain specialists, brokers, building society administrators) who have the ability to allot scant resources, especially lodging, among 'contending individuals. Since, for a great many people, purchasing a house requires getting credit from a bank or building society, the branch directors of these foundations are set in a place of force. They control, figuratively speaking, the 'door' to house purchasing, shutting it to the individuals who are considered to have a helpless fico assessment. In the USA bequest specialists have become watchmen by guiding black purchasers to specific local locations and away from others. The property managers of secretly leased convenience are additionally ready to play out a similar job of discrimination.

Gatt (General Agreement on Tariffs and Trade)

This was set up in 1947 by the world's driving countries after the second world war to advance streamlined commerce by eliminating taxes and different kinds of exchange hindrance. The first 23 signatory countries have now developed to almost 100. Since the mid-1980s, quite a bit of its business has been to determine the trade friction between the EU and the USA over agrarian endowments. An essential shortcoming of Gatt is that every part state claims all authority to protectionism when it accepts that its exchanging advantages are as a rule unfavorably influenced. As of late, it has been supplanted by the world trade organization (WTO).

GDP (Gross Domestic Product)

The absolute worth of labor and products created by the economy of a country over a predetermined period, ordinarily a year. As far as worldwide examinations, it is more suitable to communicate GDP in per capita terms. Rate commitment to GDP is a generally utilized strategy for checking patterns in the sectoral equilibrium of the economy, for example, evaluating the changing commitments made by the four economic sectors to the public monetary exertion.

Gender

The differentiation among sexual orientation and sex is progressively being made. While the male and female genders are organic realities, sexual orientation is molded by society and culture. It is to do with the jobs that society anticipates that men and women should play. The conventional generalization was for men to be the 'providers' and for ladies to be kid carriers and home-attendants. That view is as yet held in numerous pieces of the world and built up by religions like Islam. Be that as it may, in different regions, sexual orientation jobs are being changed, as confirmed by more ladies working external the home and the approach of 'house-fathers.'

Gentrification

An interaction happening in inner-city regions whereby old, inadequate lodging is purchased, modernized and involved by working class and rich families. The cycle is all around showed in internal London locale like Chelsea, Fulham and isling ton, which have gotten a lot of pursued and costly private areas. Improvement is likely set off by the endurance of once-rich, however once-over lodging and by locational benefits like prepared admittance to focal city work and administrations. It is likewise presumably helped by the accessibility of progress awards. When begun, the interaction is most likely supported by the apparent social eminence got from living in such in vogue regions. Improvement addresses a fascinating inversion of the ordinary filtering measure, in that it's anything but a social updating of old local locations.

Geography

The substance, extension and accentuations of geology have gone through significant change in the course of recent years, as worldview has succeeded worldview, and it is profoundly impossible that any one meaning of the subject would fulfill everybody. Most are concurred that it involves investigation of the earth's surface as the home of humanity. Be that as it may, how much topography is the study of spatial distributions and spatial relationships, how far it is worried about the cooperation among individuals and their actual environment, and how much investigation of the region is the focal point of the subject – these are largely matters for banter. As of late, the center appears to have moved towards an additional issues-based methodology, going in scale from to the nearby to the worldwide, from pollution to destitution the way that topography is situated at the interface between the normal and sociologies exacerbates showing up at a general definition.

Ghetto

Initially the term was utilized to indicate that piece of a town or city held for the Jewish people group, however these days it alludes to any local location that is generally involved by one ethnic or social minority bunch. The rise of a ghetto addresses a level of segregation achieved halfway in light of the fact that individuals from a minority bunch wish to live respectively and mostly because of prejudicial pressing factor by the host local area.

While numerous ghettos are inner-city spaces of helpless lodging, they are not only so. Where ethnic gatherings have had the option to climb the social scale, they have moved away from the slums and get moved in spaces of better private quality, accordingly framing what have been called plated ghettos. Albeit the ghetto is migrated, it is protected – it may even extend. Similarly, a ghetto may step by step disseminate because of assimilation.

Individuals from the ethnic gathering will in any case freely partner in what are named metropolitan towns.

Geothermal Heat

The warmth energy got from hot rocks inside the earths outside layer. Geothermal warmth breaks to the surface via heated water or steam, or is gradually directed upwards through the actual stones. Geothermal warmth is adequate to cause some basal liquefying of ice sheets and ice sheets, and in this way adds to their sliding over bedrock. Geothermal warmth can be utilized in the age of electricity – for instance, in Iceland, where geothermal warmth supplies the country with practically the entirety of its energy.

Geyser

An irregular wellspring of high temp water and steam, erupting from a little opening in the earth's outside layer. The water is warmed up, geothermally, in supplies underneath the surface. These are associated with the power source by a tight line. Temperatures inside the repositories and lower pipe are raised to above 100°C and the warmed water starts to extend. The water in the upper line is lifted towards the surface and, unexpectedly, as pressing factor is 'delivered' in the lower pipe, some portion of the water is out of nowhere changed over into superheated steam. As the last grows it discharges the water in the line into the air, regularly in dynamite style.

GIS (Geographic Information System)

A store of topographical information held in computerized structure on a pc. The limit of the cutting-edge pc is with the end goal that huge measures of data can undoubtedly be put away, refreshed and broke down. The development of GIS is firmly identified with remote sensing, which is continually giving new data about the outside of the earth and different planets.

Glacial Diversion of Drainage

The change of the preglacial waste example coming about because of the exercises of an ice sheet or glacial mass. Such redirections might be impermanent (enduring during the frosty scene just) or – where the redirected stream channels are profoundly chiseled into rock – may get lasting. The components of waste redirection change. The development of ice-negligible lakes (proglacial lake) may prompt the arrangement of overflow channels.

In any case, it is presently accepted that some purported flood channels were really created by subglacial surges of meltwater, streaming at incredible hydrostatic pressing factor underneath rotting ice sheets and equipped for cutting profound channels across old watersheds. Lasting seepage redirection may likewise be related with glacial watershed breaching.

Glacial Stairway

The unpredictable long-profile of a glaciated valley, coming about because of the arrangement of substitute stone bowls and rock steps. Some frosty flights of stairs are topographically resolved; the bowls result from cold over deepening including abrasion and culling of the more fragile rocks, while the means are created from harder, more gigantic stone outcrops. Frosty flights of stairs are believed to be identified with the idea of glacier flow.

Along huge valleys there are regularly discovered exchanging segments of compressing and extending flow, maybe coming about because of little starting breaks in the valley long-profile. The areas of packing stream produce chilly over deepening, while at the same time expanding stream is related with decreased erosion. Subsequently the underlying abnormality of the profile turns out to be always overstated, and a common icy flight of stairs is shaped.

Figure 24. Stairway glacier.

Source: Image by flickr.com.

Glacial Watershed Breaching

An interaction that happens in its easiest structure when the power source of ice from a gathering of cirques and additionally a valley is hindered by a bigger ice sheet. The ice is compelled to develop in level, and may in the long run escape across a previous watershed, following a line of shortcoming, (e.g., a preglacial col). The last perhaps so extended by exceptional frosty erosion that it's anything but a course for postglacial streams.

Glacier Flow

The mind-boggling measures by which ice is moved from the accumulation zone of the icy mass, through the ablation zone, to the nose. There are three significant systems, (i) basal sliding – where ice temperatures are at, or near, 0°C a layer of basal meltwater promptly shapes among ice and bedrock, in this manner going about as an oil and supporting the sliding interaction. In numerous calm glacial masses, basal sliding comprises somewhere in the range of half and 90% of complete icy mass development.

Notwithstanding, in 'cold-based' or polar ice (where ice temperatures might be just about as low as 10–20°C underneath freezing) basal sliding might be thoroughly missing, (ii) internal disfigurement – ice gems under tension are influenced by recrystallization and consistent slight developments comparative with one another. In this manner the ice all in all seems to take on 'plastic' characteristics. Inward misshapening is generally more significant, as a part of all out glacial mass stream, in 'cold' than in 'warm' ice. (iii) faulting inside icy masses – this may happen where the glacial mass is emphatically packed, as at the foundation of steep ice falls or at the nose where the glacial mass is superseding terminal moraine. At such focuses thrust faults may create, with the ice lying over the issue progressing over the less portable ice underneath.

Gley Soil

A dirt created under states of discontinuous waterlogging, bringing about obstructed soil seepage (as in valley bottoms or where a dish has framed in the B-horizon). In an anaerobic climate (e.g., one not exposed to air circulation) oxidation of ferric mixtures is confined, and the dirt fosters a somewhat blue dark mottled appearance attributable to the presence of ferrous mixtures. Where pockets of oxygen are available, oxidation may happen to frame red or orange blotches. Gley soils will in general be tacky, minimal and show no conspicuous soil structure.

Global City

One of the world's driving urban communities – a significant node in the complex monetary organizations being delivered by economic globalization. The impact of worldwide urban communities (e.g., London, New York, Tokyo) is connected to their arrangement of monetary and maker administrations. The majority of the main TNCS have their base camp there. Worldwide urban communities have amazing focal regions that represent their huge influence, status and riches. They have been depicted as the 'order capitals' of the global economy. As a result of their scale, they are hefty shoppers of resources and enormous makers of pollution. In their public settings, worldwide urban communities make gigantic levels of spatial dissimilarity and imbalance. Each addresses an enormous polarization of venture and business. Many are really primate cities.

Global Economy

The developing macroeconomic framework that inexorably connects together every one of the nations of the world. It is to a great extent to do with the overall abuse of resources, and the creation and promoting of labor and products.

Globalization

Any cycle of progress working at a world scale and having overall impacts. It tends to be physical (e.g., ocean level changes) or human (e.g., financial development) or both (e.g., an unnatural weather change). The cycles are basically topographical in light of the fact that they can possibly influence a significant part of the earth's current circumstance and individuals. They are additionally topographical in light of the fact that they can prompt various results in various pieces of the world. It isn't right to believe that globalization is just to do with the development of the global economy.

Figure 25. Due to globalization, the world is becoming highly interconnected.

Source: Image by Pixabay.com.

Global Shift

This term is normally restricted to the rearrangement of assembling at a worldwide scale. Following the industrial revolution, a large part of the world's creation got moved in America and Europe. During the 1970s and 1980s, be that as it may, TNCS started to move assembling to the NICS and supported areas in the third world. This shift was driven by the quest for less expensive and less unionized work, less expensive land and less controls on contamination.

It was made conceivable by propels in transport that brought down the expenses of moving crude materials and completed items over huge distances. The moving proceeds right up 'til today. Afterward, the previous fortresses of assembling have either gone through de-industrialization or modern restructuring. Their economies have likewise encountered another 'shift' as they have gotten more subject to the tertiary and quaternary sectors.

Global Village

A term used to pass on the possibility that individuals all throughout the planet are being moved nearer together into one worldwide local area. Advances in transport have diminished the time and cost associated with defeating the friction of distance. Advances in communications are likewise having a significant impact through their capacity to keep far off areas in consistent and quick touch. The term global city is likewise used to portray a similar turn of events, however furthermore it underlines the significant point that rising urbanization is a basic piece of this 'meeting up.' The rise of the worldwide town and the worldwide city are significant results of globalization.

Global Warming

The right now held view that the earth's temperatures are encountering a huge, assuming sluggish, ascent, as demonstrated by: a wide scope of environment information; proof, for example, the current downturn of numerous glacial masses and ice covers, and of logical pointers in exhausts from ice covers. Utilizing pc created models, it has been anticipated that by the center of the following century, worldwide temperatures may ascend on normal by 5°C or more (basically in polar locales), with subsequent relocation of the earth's climatic belts and genuine ramifications for grain-delivering regions like the prairies of America.

Another probably impact will be an ascent in ocean level of 2–3 m or more, with heartbreaking ramifications for some low-lying locales (undoubtedly, numerous pacific islands would be completely immersed). Numerous researchers accept that an earth-wide temperature boost is a human-incited wonder, coming about because of the development in the climate of inordinate amounts of carbon dioxide (from the utilization of petroleum products, and because of the annihilation of rainforests in Amazonia and other tropical districts).

This improves the normal greenhouse effect in the earth's environment. In any case, there is a lot of debate about the degree, causes and likely effect of an earth-wide temperature boost. An idealistic view would be that the earth as a characteristic framework is, over the long haul, equipped for acclimating to these changes.

In the present moment in any event, most governments and seaside specialists are viewing the danger appropriately as they plan for what's to come.

GNP (Gross National Product)

This is the GDP of a country in addition to all the pay procured by inhabitants from venture abroad, yet less the pay acquired in the country by outsiders abroad. Likewise, with GDP, the action is of most noteworthy worth when it is communicated in per capita terms. It is broadly utilized, as when making worldwide examinations of financial development.

Gondwanaland

The name given to a previous huge land-mass, including the southern mainland of America, Africa, Asia, Australia and antarctica. During the Mesozoic time, this 'supercontinent' started to part, and the different segments floated during the tertiary period to their present-day areas.

H

Hadley Cell

A component of the earth's overall dissemination in low scopes, proposed by Hadley in 1735. Serious warming of the surface in the equatorial zone causes largescale inspire of air.

This is repaid by a low-level stream towards the equator from the sub-tropical anticyclone at about 30°N. This is itself framed part of the way from the plunge of air that has moved at a significant level away from the equatorial zone. The surface breezes in the resultant 'cell' are redirected by the earth's pivot to bring about the ne trades. An equivalent cell is recognizable in the southern hemisphere. The earth's overall course contains, notwithstanding the Hadley cells, two different cells in every half of the globe: a northerly low-level stream from the sub-tropical high towards the polar front (with a repaying return stream at an undeniable level); and a southerly stream from the polar anticyclone (and a remunerating return stream at high heights).

In later occasions this straightforward model of the earth's overall flow has been changed to assess jet streams and flat examples of air development, (e.g., moving high-and low-pressure frameworks close to the surface).

Halophyte

A plant that flourishes in a salt impregnated soil, as in an estuary or on a beach front salt marsh, or within the sight of a pungent splash (as on the lower portions of ocean precipices). On salt bogs halophytes play out the significant undertaking of catching sediment when immersed at elevated tide, in this way making the swamp develop upwards after some time. In this regard they can be viewed as a pioneer species, the primary stage in a vegetation succession.

Hanging Valley

A feeder valley that joins a primary valley via a sharp break of incline, frequently set apart by a waterfall. The most successive reason for hanging valleys is frosty erosion. An enormous glacial mass in a primary valley will cause over deepening in overabundance of that accomplished by a more modest glacial mass in the feeder valley. At the point when the glacial masses have softened, the feeder valley is left 'hanging' for however much many meters over the fundamental valley floor. At the point when a waterway gets back to the feeder valley, it cuts into the break of slant and, on schedule, the cascade will be supplanted by a progression of rapids inside a precarious sided gorge. An alluvial fan might be shaped where the feeder waterway joins the primary stream valley, and the quick decrease in inclination causes a drop in the waterway's speed prompting deposition.

Harbor

An asylum or harbor for delivery including a stretch of shielded water, near the shore and shielded from the untamed ocean by regular (e.g., a spit or headland) or fake (e.g., a sea wall or breakwater) implies. Rias, with their profound and segregated channels, presumably give probably the best normal harbors, like those in Sydney, Australia, and Milford haven, Wales.

Harmattan

An exceptionally dry wind, blowing from the Sahara Desert towards the Gulf of Guinea in Africa. It is prevailing from November January, when the high-pressure cell over the desert is most firmly evolved. The harmattan is described by very low relative humidity (some of the time of under 10% at late morning) and cool temperatures. The breeze is regularly dust-loaded, up to statures of 3000–5000 m, and perceivability might be decreased to 300 m or less.

It's anything but a time of alleviation from the predominant sticky warmth of Africa. Be that as it may, its belongings are not unexpected a long way from big-hearted. The beginning of the breeze is set apart by an increment in respiratory contaminations, and germs borne by the residue particles give off an impression of being answerable for episodes of cerebral spinal meningitis. The fire danger, because of the outrageous parching, can be serious.

Figure 26. Boa-vista harmattan.

Source: Image by Wikimedia.

HDI (Human Development Index)

Utilized as a proportion of development in a country and for making worldwide examinations. The list considers three factors that are given equivalent weighting: pay per capita; grown-up proficiency; future. The HDI takes the most noteworthy and least qualities recorded for every factor by the nations of the world. The span between them is given a worth of 1 and the worth of every nation is then scored on a size of 0 to 1 (from most exceedingly terrible to best). The HDI is the normal score of the three factors thus, as well, is communicated as a worth somewhere in the range of 0 and 1. The most well off MEDC have a record drawing nearer 0.999 and the least fortunate LEDC range down to under 0.300.

Head

A store of inadequately arranged stones and boulders found at the foot of inclines and sea cliffs, and coming about because of past periglacial action. The flotsam and jetsam, every now and again precise, is delivered by freeze-thaw weathering, and then solifluction takes place down the slant or bluff face to gather as a basal store, regularly a few meters in thickness.

Headward Erosion

A cycle regularly connected with steepened areas of a waterway's long-profile, whereby disintegration happens in an up-valley heading. One normal illustration of headward disintegration is the upstream downturn of waterfalls. Headward disintegration is likewise normal at springs (e.g., in chalk and limestone) where it is all the more explicitly alluded to as spring-sapping.

Heat Balance

The normal state of equilibrium in the atmosphere between approaching sun-oriented radiation and active warmth misfortunes because of reflection or re-radiation from the surface. The warmth balance along these lines forestalls, in any event for the time being, critical ascents or falls in climatic temperatures. Be that as it may, the warmth balance isn't kept up, in the terms expressed, at each point on the earth. For instance, at or close to the equator more warmth is acquired from sun-based radiation than is lost by re-radiation, though in higher scopes more warmth is lost than is acquired. In this way, equilibrium is kept up by the horizontal exchange of warmth, graciousness of moving air masses and sea flows that convey heat from warm to cold scopes.

Heat Island

The wonder whereby temperatures inside a city or huge town are frequently essentially higher than those of encompassing rural regions. The warmth island impact is most noteworthy around evening time, when city temperatures might be 6–8°C higher than those in country regions, as warmth put away by urban surfaces during the earlier day is gradually delivered. There are different components included, for example, the arrival of warmth by homegrown and modern fuel utilization; the high limit with respect to warm assimilation of numerous metropolitan surfaces, for example, block facades and black-top streets; the diminished utilization of warmth by cycles like vanishing, attributable to the deficiency of water by quick waste and the resultant 'dry' climate; barometrical contamination, which lessens misfortunes of long-wave radiation starting from the earliest stage night. It has been shown that the focal point of London has a mean yearly temperature of 11.0°C, contrasted and 10.3°C for the suburbs and 9.6°C in provincial regions. At 9pm on 26 February 1988, the focal point of oxford recorded a temperature of 4°C contrasted with about 0.5°C in the field only 7 km away.

Figure 27. Urban heat island of Chicago.

Source: Image by flickr.com.

Heath (Heathland)

A space of generally treeless nation, overwhelmed by heathers, gorse and different grasses ready to flourish with poor acidic soils. Genuine heaths (e.g., those of the new forest and Dorset in s England) are restricted fundamentally somewhere in the range of 50° and 60°N, and are normally under 200 m above ocean level. Heaths have been framed by the degeneration of previous forest regions, because of the far and wide consumption of trees by neolithic man, for development and the feeding of creatures. This has been joined by soil impoverishment (by dynamic leaching, particularly in sandy regions), fermentation, and the development normally of a dainty surface peat layer. Right now, heaths, which give an extraordinary climate to specific plants and creatures, are by and large truly decreased in degree by urbanization, afforestation, and reclamation.

Heavy Industry

Basic industry of public financial significance and in which huge amounts of materials are dealt with or handled; for example, coalmining, iron and steel creation, shipbuilding, petrochemicals.

Helicoidal (Helical) Flow

A kind of stream related with crooked, and specifically wandering, waterway channels. It is felt that the primary momentum of the waterway will in general stream in an orderly fashion, along these lines impinging on the external bank where the channel twists. Therefore, a slight head of water will be developed, prompting a compensatory return stream across the channel. Since the most extreme speed in a stream (the thalweg) is simply underneath the water surface, this compensatory stream will be near the channel floor, giving a roundabout movement when seen in cross-segment. This optional stream is superimposed on the principle downstream progression of the waterway, bringing about a kind of winding movement, or helicoidal stream. It is conceivable that helicoidal stream has some influence in the transaction of sediment, disintegrated from the external banks of meanders, to within banks where it is stored.

HEP (Hydro-Electric Power)

Electric force created in turbines, the intention energy for which is gotten from moving water. At times, HEP is produced at regular waterfalls (e.g., Niagara), in others it's anything but a dam across a valley (e.g., terrific coulee dam, USA), while it can likewise be accomplished by funneling water down a mountainside. In every one of the three occasions, the basic prerequisite is a decent head of water kept up at a steady pace of supply.

Heritage Coast

An assignment made in Britain since 1970 to guarantee that stretches of bank of high picturesque quality are shielded from advancement and are dependent upon successful public administration.

Heritage Tourism

The advertising of scenes, especially their strange beautiful characteristics and noteworthy relics (e.g., dignified homes, archeological locales, old metropolitan regions, and so on) as vacation spots. A type of tourism that looks to abuse the past and the novel as resources. Once in a while alluded to as the legacy business.

Heterotroph

An organic entity unequipped for self-taking care of, which infers its important stockpile of sugars and starches from autotrophs, for example, a cow that eats grasses. Heterotroph's address

High-Technology Industry

Every now and again alluded to as 'cutting edge industry,' a mechanical action including trend setting innovation and initiating the fifth Kondratieff cycle. The term covers a significant variety of exercises, like the production of semi-conveyors, pcs, computer chips and modern robots; the advancement of new metals and pottery; media transmission and data innovation; biotechnology and fiber optics. A crucial contribution to all innovative industry is R&D, frequently embraced at close by colleges, specialized schools and other public establishments. In this manner, admittance to R&D foundations can establish a huge area factor for these new businesses, as does nearness to an air terminal to guarantee fast admittance to public and worldwide business sectors.

Hill Farming

General cultivating did in upland regions, however in Britain the term is rigorously applied, for government award purposes, to a particular class of cultivating. It is especially worried about sheep-raising and the raising of meat steers.

Figure 28. Hill farming.

Source: Image by pxhere.com.

Honeypots

A term every now and again utilized with regards to recreation and tourism to indicate spots of extraordinary interest or allure that are profoundly well known with guests and that will in general turn out to be exceptionally clogged at busy times, for example, Land's End, the lake district, the tower of London. With regards to making arrangements for diversion and the travel industry, honeypot approaches have frequently been received (particularly in public stops) that support the advancement of chose honeypots determined to guarantee: that legitimate arrangement is made at these locales for the convenience of guest traffic (giving vehicle leaves, excursion destinations, latrines, and so forth); redirecting traffic from rather more 'delicate' and pristine regions that couldn't in any way, shape or form adapt to huge quantities of guests without genuinely reducing their natural allure.

Horizontal Extension

One of three manners by which an endeavor may extend including the takeover by a firm of other, apparently rival, organizations in a similar business. Such assimilation of different firms gives the chance to expand the size of creation or activity, and hence for the augmented venture to profit with economies of scale.

Human Ecology

The utilization of the ideas of ecology to examination of the connections among individuals and their environment. It is to be found in burgess' concentric zone model of city structure.

Human Geography

That piece of geography worried about the spatial investigation of the human populace. It envelops such parts of the populace as numbers, piece, monetary and social exercises, and settlement, with every viewpoint having a tendency to produce its own part of human geology.

Human Rights

The un universal declaration of human rights (1948) states that all individuals reserve the option to: life, freedom and instruction; opportunity of development, religion, affiliation and data; an ethnicity; correspondence under the watchful eye of the law.

Humus

Natural material inside the soil got from leaves, twigs, roots, dead living beings, creature excreta, and so forth, which has bit by bit been separated by compound rot and the exercises of miniature organic entities into a finely isolated, dull brown thick substance. Inside the dirt layers (especially the a-horizon) the humus goes into a mind-boggling compound relationship with clay minerals, shaping the alleged clay-humus complex or colloid. In cool, clammy environments, natural matter may amass at the dirt surface as a crude humus layer.

In coniferous woods and heathlands, plant litter is made fundamentally out of mixtures that rot gradually, giving supposed mor humus. This is firmly acidic, debilitate soil living beings, and is related with moderate humification (the rot and joining of humus into the dirt). Paradoxically, mild deciduous forests, with broad harvest time leaf fall, give ponder humus. This is all the more promptly decayed and is wealthy in soil organic entities, particularly night crawlers, which help to blend the humus and soil particles; it is additionally connected with base-rich soils.

Hurricane

An exceptionally incredible hurricane, described by winds of outrageous speed (more than 120 km/hr) and fit for causing boundless harm ashore, just as establishing a genuine peril to delivery. Tropical storms at first structure predominantly in scopes 5–10°N and s over the western segments of the Atlantic (especially the west indies and Gulf of Mexico district), the Indian and pacific oceans. They are more continuous in the northern hemisphere (yearly recurrence 50–60) than in the southern hemisphere (yearly recurrence 20–30).

The tropical storm season in the Atlantic endures from July to October, and happens when the central low-pressure zone is uprooted northwards. Every individual typhoon ordinarily endures from a few days. The tempests are produced over warm sea waters (with surface temperatures above 27°C), and the energy 'driving' them is gotten from the huge scope condensation of clammy air and the resultant arrival of a lot of inactive warmth. For additional subtleties of pressing factor conditions, winds, climate and track.

Figure 29. Graphical representation of Hurricane.

Source: Image by flickr.com.

Hydration

An interaction of chemical weathering whereby certain stone minerals take up (retain) water. In the process changes in volume happen, actual burdens are set up inside the stone and, therefore, actual breaking down can happen, normally as limited scale chipping from the stone surface.

Hydraulic Geometry

An investigation of the changing calculation of a stream channel, communicated in boundaries like width, profundity and slant, and according to varieties in discharge.

Hydraulic Ratio

The proportion between the cross-sectional space of a waterway channel (d x w) and the length of the wetted edge (the line of contact between the water and the channel). Subsequently r = a/p, where r is the pressure driven span, an is cross-sectional region, and is the wetted edge. Pressure driven range is a proportion of the productivity of a stream, alluding to the corresponding misfortunes of energy by grinding between the streaming water and the channel bed and banks, as contrasted and the misfortunes inside the water. Huge qualities for rare related with streams with enormous discharges, and with cross-segments that are roughly semi-roundabout. Then again, low qualities (and hence decreased effectiveness) are given by little streams with a significant width depth proportion.

I

Ice Age

A geological period of widespread glacial activity, when mainland ice sheets created and glacial erosion, transport and deposition worked for a huge scope. The latest ice age was that of the Pleistocene time frame. This started exactly 2 million years prior. Albeit a significant withdrawal of the ice happened exactly 10,000 years bp, it is far-fetched that the Pleistocene ice age has really ended. Inside the actual Pleistocene, numerous significant variances of environment occurred, giving substituting frosty and interglacial periods. It was once in the past accepted that there were four primary chilly periods, yet later proof from sea floor centers has demonstrated the chance of up to 20 cold periods (each of as long as 100,000 years in span), isolated by brief interglacial of approximately 10,000 years.

Ice Cap

A constant space of land ice that is less broad than an ice sheet. Mountain ice covers involve upland plateaus, and achieve thicknesses more than 500 m. At their edges quickly streaming outlet glacial masses cause significant analyzation of the level edges. Swamp ice covers are found in spaces of delicate relief at lower heights in the high arctic.

Ice Fall

A precarious segment of a valley ice sheet, including various profound crevasses and, at the highest point of the ice fall, broken ice masses and zeniths (seracs). Ice falls happen at the top of a frosty box, where feeder ice sheets from cirque bowls unite, or where the glacial mass disregards a bar of safe stone. Stream speeds on ice falls are high, potentially in the request for 1000 to 2000 m/yr.

Ice Sheet

A broad, consistent space of land ice that may, at its edges, level into ocean ice. Just two ice sheets exist as of now, those of antarctica (covering some 11.5 x 106 km^3) and Greenland (1.7 x 106 km^3). Nonetheless, in the new past mainland ice sheets likewise covered quite a bit of new Europe and America (the Pleistocene ice sheets). The ice sheets of antarctica and Greenland are described by incredible thicknesses of ice (2000 m on normal in the previous), low temperatures (with the outcome that even the basal ice is locally at –30°C), and moderately low paces of amassing and ablation (and thusly low speeds).

Ice Wedge

A tightening mass of ground ice, shaped in a crevice a few meters inside and out under periglacial conditions. Where there is extreme chilly, the soil or regolith will agreement to shape a polygonal organization of profound breaks (gap polygons). In summer these may get involved by meltwater, which in the accompanying winter will freeze, extend and grow the gaps along the side, leading to ice-wedge polygons.

Iceberg

A huge mass of skimming ice that has split away from an ice sheet or glacial mass ending in water. The main wellspring of chunks of ice is the ross ice shelf in antarctica. As the ice, moving out from the Antarctic ice sheet crosses the ocean bed, it from the start remains grounded; nonetheless, the ocean ultimately turns out to be adequately profound to coast the 300–400 m layer of ice, prompting the splitting away of numerous plain ice masses past the 'establishing line.' The icy masses then, at that point float northwards, at a pace of a few km daily, and can infiltrate similar to 60°S in the Pacific Ocean. In the northern hemisphere numerous ice shelves are gotten from the calving of ice at icy mass noses – for instance, along the shore of Alaska.

Figure 30. Antarctica ice berg.

Source: Image by Pixabay.com.

Igneous Rock

A sort of rock framed by the cementing of magma, either inside the earth's outside (intrusive rock) or at the surface (extrusive (volcanic) rock). Volcanic rocks shift as per: compound arrangement (regardless of whether the magma is silica rich, as in granite, or essential, as in basalt, dolerite and gabbro); the pace of cooling, which is fast at the surface, giving fine-crystalled or lustrous shakes like basalt, and much slower at extraordinary profundities, giving coarse-crystalled shakes like stone or gabbro). Some molten rocks create from sedimentary rocks that have gone through a middle phase of incredible metamorphism. The last interaction might be outrageous to the point that magma is 'reproduced,' as in the emplacement of huge rock batholiths.

Illuviation

The affidavit of soil matter (natural humus, clay-size particles, and solutes) in an illuvial B-horizon, having been eliminated by eluviation from the a-horizon. Convergences of mud particles may bring about the development of a claypan, and centralizations of iron may shape an iron pan.

IMF (International Monetary Fund)

This was set up in 1944 as an office of the uno. It exists to advance global money related participation, the extension of world exchange and trade dependability. It additionally expects to kill unfamiliar trade limitations. The IMF isn't a bank as in, rigorously talking, it doesn't loan cash, yet it helps part nations to gain unfamiliar money.

Immature Soil

A 'youthful' soil in which soil-shaping cycles have had little impact. The mineral substance of youthful soils is high, the humus content is low, and there is practically zero advancement of soil horizons.

Immigration

The demonstration of moving into a space determined to get comfortable it. The term is most ordinarily utilized when alluding to the development of individuals across public frontiers.

Imperfect Competition

A market circumstance in which neither total monopoly nor perfect competition wins; a circumstance nearest to genuine as a rule. It is described by the capacity of dealers to impact interest by item marking and promoting, by limiting the section of rivalry through prohibitive practices, by the presence of uncertainty and the nonattendance, to changing degrees, of value contest.

Imperialism

The arrangement of making a domain whereby an amazing state fosters a relationship (e.g., through trade, military insurance, specialized aid) with other and subordinate domains, as previously among Britain and her states and between the former Soviet Union (FSU) and her reliant nations in Europe Dominion and colonialism are at times taken as being equivalent.

Impermeable

A term portraying rocks or shallow stores that don't permit water to go through them. This is expected either to the absence of interconnection between pores, or the shortfall of joints and bedding planes (as in certain kinds of massive stone, including dolomitic limestones). Impermeable rocks are portrayed by supporting surface waste, and paces of fluvial erosion are moderately high, albeit outright measures of disintegration will rely upon different elements, for example, rock type.

Impervious

A term that is now and then viewed as inseparable from impermeable, but at the same time is utilized to signify rocks that don't have joint planes or gaps along which water development can happen. At the end of the day a stone might be non-permeable and impermeable (like an individual square of granite), and furthermore impenetrable on the grounds that there are no interconnected breaks (as in an enormous gabbro).

Import

A decent or administration that is brought into a country from another with regards to global trade.

Import Control

Government-made gadgets intended to restrict the measure of imports entering a nation thus guarantee a more positive balance of trade.

Import Penetration

How much a nation is subject to imports. For instance, the UK has a high import entrance in engine vehicles in that a considerable lot of the vehicles sold there are made abroad.

Import Substitution

The interaction whereby a nation looks to create for itself labor and products that it once imported. Thus, import replacement ventures (ISIS).

Improved Land

A space of land that has been made more useful, as by waste, reclamation or the use of fertilizers. Utilization of the term is normally limited to the setting of agriculture.

Incised Meander

An overall term for a meandering stream that has chopped down into its anything but an aftereffect of rejuvenation, bringing about the wanders being admirably beneath the level of the nearby land. In the event that restoration is moderately sluggish, the wander will keep on dissolving horizontally as it down cuts, so framing 'another' valley that is awry in profile – this is named an ingrown wander. Genuine models are found along the seine valley among Paris and le Havre. Conversely, if downcutting is fast, a smaller valley with balanced sides is bound to frame. This is known as a settled-in wander. Incredible models are found in the valley of the wear at Durham.

Indicator Species

A type of plant or creature that has specific natural prerequisites, and would thus be able to be utilized as a marker of the condition or nature of a climate. These necessities may identify with climatic or soil qualities, or to the non-contamination of the climate. An illustration of a pointer species is the peregrine hawk, a hunter of pigeons, pigeons and other bird species, which encountered a genuine decrease in Britain from 700 reproducing sets before 1939 to likely less than 300 by the mid-1960s.

This fall was related with the passing of grown-up birds, and furthermore an inability to raise effectively attributable to thin eggshells, barren eggs or the powerlessness to lay eggs by any stretch of the imagination. Exploration showed that the reason was natural contamination, including the utilization of bug sprays like DDT and other chlorinated hydrocarbons, which were consumed by the peregrines through the prey species. Since the forbidding of the utilization of DDT, and with the guide of unbending insurance, peregrine numbers have recuperated terrifically to well more than 1000 sets, subsequently demonstrating a re-visitation of 'cleaner' natural conditions.

Indices of Segregation

The term given to a scope of various factual measures utilized in social geography to gauge the level of social separation and private isolation inside a given populace of these, four have end up being especially generally utilized:

i. the index of dissimilarity, utilized where the focal point of examination is on the rate dissemination of two explicit gatherings;
ii. the index of residential segregation-, which demonstrates the rate distinction between the circulation of one gathering and the dispersion of the remainder of the populace (list esteems range from 0 to 100 – the higher the worth, the more noteworthy the isolation of that specific gathering);
iii. the location quotient and index of similarity may likewise be utilized to evaluate the overall centralization of a specific gathering inside a given region;
iv. the Lorenz curve, which gives a graphical technique to showing isolation.

Industrial Archaeology

The investigation of the remaining parts of past mechanical movement – as, for instance, of old plants, dead mines and purifying works, neglected trenches and rail route lines. As of now, there is a lot of interest in protecting such relics and advancing them as a feature of heritage tourism. The iron bridge gorge museum is a genuine model, being particularly significant as the origination of the industrial revolution.

Industrial Complex

A huge convergence of assembling in a generally little region. Firms situated in a particularly mind-boggling advantage from agglomeration economies, especially most definitely. Such advantages incorporate great transport and communications, and admittance to a wide scope of help administrations.

Industrial Geography

The investigation of the spatial game plan and association of mechanical action. As a sub-part of economic geography, it is chiefly worried about secondary sector or assembling movement.

Industrial Inertia

The propensity for an industry to stay situated in a space when the purposes behind it first situating there could be not, at this point huge or presently don't exist. Models are when wellsprings of fuel, mineral or other raw materials become depleted in one area, or when they can be acquired all the more monetarily somewhere else. Inactivity is supported in various manners, concerning model by fixed costs (capital interest in framework), by the presence of tight linkages with different exercises situated nearby, or by the persistence of a pool of explicitly talented work. Once in a while alluded to as locational inactivity.

Industrial Location Theory

A part of industrial geography that came to unmistakable quality during the 1960s, conceivably animated by a whole lot sooner hypothesis of modern area propounded by weber. Crafted by weber and different scholars, like hoover, isard and losch, regularly make presumptions that are not really legitimate. Consequcntly, mechanical area hypothesis as of now gets little help, being supplanted by imperfect, conduct and primary approaches.

Industrialization

The cycle whereby mechanical movement (especially fabricating) accepts a more noteworthy significance in the economy of a country or district. It is generally considered as an essential element of development. While it without a doubt makes 'benefits' as monetary abundance and better expectations of living, it is important to remember a portion of its expenses, like environmental pollution, biological harm, depletion of non-inexhaustible resources and inordinate dependence on immense supplies of energy.

For a significant part of the twentieth century, it was accepted that industrialization offered the best way for LEDC to continue to invigorate the development interaction. Industrialization is currently starting to happen in a portion of those nations, as TNC quick to abuse the immense stock of modest work set up branch plants. This may look encouraging; however, it stays not yet clear whether this type of industrialization will carry enduring advantages to individuals of those nations. There is genuine worry that it will prompt an unsatisfactory level of financial and political dependence on 'unfamiliar' nations and organizations.

Industry

In its amplest sense, any work or action embraced for acquire and that advances business (e.g., horticulture, manufacturing, retailing, tourism). At times the term is utilized explicitly to mean essentially the assembling of merchandise (e.g., as customarily in industrial geography). In Britain, the public authority makes the differentiation between manufacturing industry and essential industry, the last involving mining, quarrying, public utilities, transport and correspondences, agribusiness, fishing and ranger service. Other potential qualifications incorporate the separation of light industry and heavy industry, or of extractive industry, manufacturing industry and service industry.

Infant Mortality

The normal number of passing of babies under 1 year old enough for every 1000 live births. High baby mortality is regularly characteristic of restricted clinical benefits, hunger and impeded development.

Inferential Statistics

Measurements used to offer expressions about a painstakingly characterized population from an appropriately chosen arbitrary example of that populace; factual methods used to test how far examples address the entire. To this end, inferential insights depend on probability hypothesis and on the utilization of significance tests. Some of the time alluded to as inductive insights, they may likewise be utilized in the testing of theories.

Infiltration

The development of water, gotten from precipitation or liquefying snow, into the soil. The pace of penetration, called the invasion limit, relies upon a few variables, for example, soil porosity, the level of compaction of the dirt surface, the presence of plant roots, and how much soil moisture is as of now present (the predecessor state of the dirt). At the point when downpour happens, penetration is from the start fast. Be that as it may, as the unfilled voids inside the dirt become loaded up with water, the pace of passage will back off to rise to the sums lost at the foundation of the dirt, or by throughflow. At the point when downpour falls at a rate surpassing invasion limit, overland flow (penetration overabundance overland stream) will happen. In any case, this is probable just in outrageous tempest conditions or in inadequately vegetated semi-bone-dry areas.

Inflation

A cycle joining rising costs and degrading cash. A condition in which the volume of buying influence in a given region is continually running in front of the output of labor and products, with the outcome that as earnings and costs rise, so the worth of cash falls.

Inflationary Spiral

A vertical development of costs that is incompletely the outcome, and mostly the reason, of expansions in wages and pay rates, profits, loan fees, and so on

Informal Sector

The piece of the economy that involves a scope of exercises. In MEDC, these are generally of a help kind and are attempted by people. There is almost no authority command over them. It incorporates exercises like homegrown cleaning, childminding, bar-tending and different types of easygoing working, where installments are made, however frequently not pronounced, consequently keeping away from installment of personal assessment and public protection commitments. It is prevalently alluded to as the dark economy, in view of its undisclosed and secret nature.

In LEDC, the casual area shows the accompanying repeating highlights: a prevalence of independently employed individuals and little family endeavors; a high frequency of youngsters in the labor force assisting with enhancing family wages; the creation of bad quality merchandise through the reusing of waste and utilization of second-rate crude materials; the showcasing of labor and products basically attempted in the city.

Infra-Red Radiation

Undetectable electromagnetic waves that are seen as warmth. In light of their frequencies, they can enter cloud and haziness. Infra-red photographic gadgets are utilized in remote sensing and can recognize highlights of the earth's surface paying little heed to light or barometrical conditions.

Infrastructure

The amenities and administrations that are fundamental to most kinds of human action. These incorporate the arrangement of streets, power supplies, interchanges, water and sewage removal frameworks. Some would characterize the above models as actual foundation, as unmistakable from social framework that incorporates such things as schools, emergency clinics and jails. Both are given at a local area, territorial or public level essentially by open assets, to which singular clients contribute to a great extent through the duty of rates, expenses and standing charges.

Initial Advantage

The benefit that accumulates to a firm, city, region or country because of it being first be it in the presentation of another item, the misuse of another resource, the business use of new innovation or in the opening up of another market. When the underlying benefit has been seized, then, at that point the interaction of cumulative causation sets in, supporting the advantages to be acquired.

Inlier

An outcrop of more seasoned stone, encircled by more youthful rocks, that has been uncovered by the limited evacuation of the more youthful rocks by erosion. Inliers are usually connected with anticlines and domed designs. Disintegration of these will uncover more seasoned rocks raised along the overlay axis.

Inner City

That piece of the built-up area near the inner-city area, extensive spaces of which are portrayed by old and old lodging, by multi-family occupation of abodes, by helpless business and social services, and by the presence of mechanical and business endeavors that add to the urban blight regular of such regions. Inner city populaces will in general include three particular gatherings:

i. native inhabitants – mostly individuals in the more established age gatherings;
ii. outsiders (regularly having a place with minority ethnic groups);
iii. drifters – individuals moving who utilize the region as an impermanent base.

The last two gatherings move in light of the fact that their monetary conditions oblige them to possess low-lease convenience, while the main gathering of individuals stay there in view of their fixed status or failure to manage the cost of better and more costly lodging somewhere else, and on the grounds that they should be near the downtown area (essentially to save money on transport costs). The inner city comprehensively harmonizes with the zone of transition in the concentric zone mod.

J

Jet Stream

At roughly 9–15 km above the Earth's surface, a high-altitude, very fast-flowing wind with speeds ranging from 180 to 270 kilometers per hour blows. There are a lot of jet streams, but one of the most significant is the Polar Front Jet, connected to the frontal area at mid-latitude. Midlatitude depressions, which are directed eastward by this system, appear to play an extremely essential role in the formation and migration of these depressions. At around 30°N, in the Northern Hemisphere, the Polar Front Jet migrates northwards, initiating a second Sub-tropical Jet Stream, which has a more continuous westerly flow, which is begun at approximately 30°N. In the summer, an Easterly Tropical Jet Stream forms across India and Africa, bringing warm air from the equator. It is most likely that jet streams occur along lines where the tropopause has a strong gradient or is broken in some way.

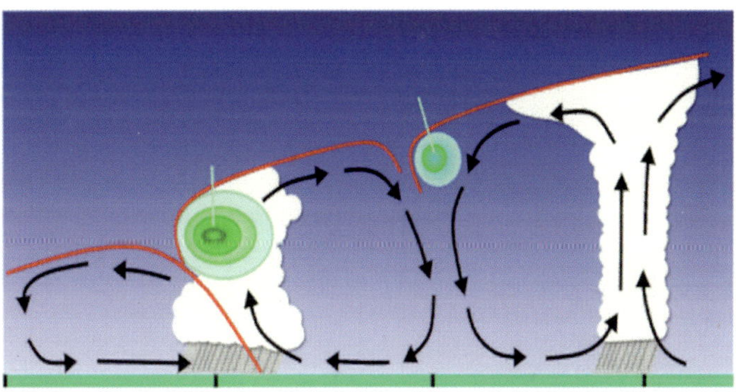

Figure 31. Diagram of Jetstream.

Sources: Image by Wikimedia.

Joint

A thin crack or split in the rock that does not move on either side of the split. Connections are created in many ways: by contraction during the cooling of igneous rocks; through earth movement pressures; through dilation; through pressure release.

Joint-Block Removal

Glacier ice causes the dissociation of blocks that are joined together. Jointed block removal, also known as plucking, is thought to be crucial in the development of cirque headwalls, the risers of rock stairs, and the down glacier or stolon faces of a Roche Moutonnee, among other things. Some believed that the ice had frozen on the granite mass and had pulled it apart from the rest of the structure. Nevertheless, this is now somewhat questionable due to the fact that the tensile strength of ice is far lower than that of rocks and, in this circumstance, the moving ice would most likely shatter first unless the rocks had already been sufficiently damaged by weathering or dilatation.

Journey-To-Work

Traveling between one's house and one's place of employment, often known as commuting the distance travelled to and from work, as well as the amount of time spent travelling, may be measured. Another significant factor in the examination of journey-to-work patterns is the mode of transportation utilized.

Just-In-Case Production

A production organization system that involves the storage of raw materials and components in a facility. This is done only in the event that the costs of these products have increased or that the demand for production in the plant is suddenly increasing. It has proven inefficient and wasteful.

Just-In-Time Production

This form of production organizations is intended to reduce the number of raw materials and component stock retained by a plant by as much as possible. It is more cost-effective in this manner. The production chain must be meticulously scheduled in order to fulfil demand forecasted accurately in order for it to function successfully. Ordering systems must be as efficient as possible. The defect-free raw materials and components must be used, and their supply must be completely trustworthy. Japanese manufacturers were the first to utilized the technique, and they continue to do so to great success. Since then, it has become widely utilized in NICS and, more recently, in the United Kingdom.

K

Kame

Gravels and sands piled high on top of one another, created by melting glaciers or glacial ice sheets melting into the ground. Kames are distinguished by their size, shape, and origin. Some are little hillocks that rise only a few meters above the surrounding terrain, whereas some are steep-sided hills that rise 30–50 meters over the surrounding terrain. Asymmetrical hillocks are formed as a result of the accumulation of sediment in large sinkholes formed by glacier ablation close to the glacier's margins.

Some kames form as a result of sediment fans or deltas laid down along the ice margin, often where small lakes are impounded, which subsequently form asymmetrical hillocks. It is expected that a complex of kames will form when the ice melts at some point. Kame terraces form between ice and a valley wall and are frequently seen within marginal lakes that have gotten encrusted with silt. When the ice melts, the ice contact slope becomes the steep terrace edge, which is formed by the melting ice.

Figure 32. Image showing kame.

Sources: Image by flickr.

Kant's Index of Concentration

The distance between habitations is used in the study of rural settlement to determine the degree of concentration or dispersion of settlement, as represented in the number of habitations.

Kaolin

Kaolinite is a white clay formed as a result of the breakdown of feldspar for example, inside granite caused either by recent chemical weathering or by hydrothermal processes acting at the time of rock formation also known as Pneumatolysis. In refined form, employed in a wide variety of products from China to cosmetics, paper to pharmaceuticals, is utilized in many manufacturing processes.

Karren

Furrows formed on the surface of limestone as a result of solutional processes acting on the rock. A number of them are produced by acidulated rainwater on exposed limestone rocks, while other types are the result of subsurface weathering. In terms of scale, Karren range from shallow grooves running down limestone faces to deeper channels that can be several meters long and up to 0.5 meter deep.

Figure 33. Image showing Karren.

Sources: Image by Wikimedia.

Karst

In the former Yugoslavia, there is a unique region of limestone. A limestone region characterized by subterranean drainage and an abundance of surface solutional formations such as dolines, uvalas, and poljes is now commonly referred to as a doline zone. When it comes to karstification, it refers to the gradual disappearance underground of surface streams by way of sink holes, the formation of underground passages and cavern systems, the lowering of the limestone water table, which results in numerous dry valleys and depressions, as well as the application of extensive surface and subsurface solution, which results in small features like Karren and larger features like as dolines.

Katabatic Wind

On a frigid night, a local breeze blows downward, bringing the temperature down. Katabatic winds are caused by the movement of dense air that has been chilled by contact with slopes that have been themselves cooled by nighttime radiation, which is particularly noticeable on clear days and nights. It is possible that frost and fog will form in valley bottoms if the cold air condenses and pools there. The occurrence of a form of katabatic wind can also occur when cold air drains from a glacier's surface into a valley underneath it.

Kettle Hole

A circular depression formed by the slow disintegration of a block of ice covered by underlying sediments, which was originally filled with meltwater and then drained. Kettles may now occasionally be found in great numbers in front of alpine glaciers, from which streams wash off and store great quantities of sand and gravel, which conceal from the glacier snout masses of calved ice. When a number of kettles come together, they can produce huge bodies of water.

Key Settlement

An abbreviation used in rural planning to designate a village or small town that has been identified for promotion as a small growth Center. According to the approach, isolated rural regions are losing population at a faster pace than they can replace it. In the designated major settlement, it is believed that growth will be fostered by concentrating rather than spreading investment, and by utilizing that investment to expand and improve the range of central place services available, as well as to provide housing and introduce new forms of employment. Eventually, this expansion will spread to neighboring areas and surrounding communities that were not initially targeted. Because of this, the main village serves as a miniature growth pole.

Kibbutz

An Israel-based rural settlement in which land is commonly owned and managed and structured according to community ideals. Early kibbutzim were mostly agricultural in nature, but more recently, there has been an increase in and success with industrial production on a larger scale.

Figure 34. Image showing Kibbutz.

Sources: Image by Wikimedia.

Kinetic Energy

The water flow, breaking waves, sliding ice, and other forms of heat friction are constantly dissipating free energy as heat friction. This produces more than doubles the kinetic energy of an equally sized river that flows at half the speed of a large, fast moving river.

Knickpoint

The point when a lengthy profile of the river is markedly steeped due either to a resistive rock outcrop or to a relative drop in marine level. The latter kind occurs as a result of the fact that the offshore gradient is often greater than the gradient of the river's lower course. As a result, a decrease in sea level will result in the addition of a somewhat steep portion to the expanded river profile.

When this happens, it will be carried upstream in the form of an erosion wave, which will frequently form a gorge as the river cuts down, ultimately reaching the head of the stream. In several rivers, a sequence of knickpoints may be seen, which are associated with the many base-level fluctuations that occurred during the Pleistocene. When a river is captured, knickpoints are produced as a result. In effect, the capturing river serves as a reduced local base-level for the caught stream to strive towards as it moves downstream.

Knock-and-Lochan

A kind of terrain seen in NW Scotland, which was due to an erosion by an ICE SHEET of a relatively low hard rock region. The second scans geological weaknesses regions and lines i.e., well-jointed rock, faults, and belts shatter, producing many depressions. These lakes or lochans and peat bogs are inhabited following the departure of the ice. The low ringed hills, typically with little soil or plant or knocks, which are searched through glacial abrasion apart from the depressions.

Knock-On Effect

In a wide range of scenarios, the implications of a specific action, decision or event are generally indicated, for example, the 'ripple' impact connected with an opening of a new freeway or the broader impact of a big steel plant's closure. In certain cases, using a multiplier effect phrase may be synonymous.

Kondratieff (Long Waves) Cycles

Kondratieff was a Russian economist, who, in 1925, published a paper stating that empirical evidence indicated that capitalistic economies are subject to business cycles for about 50 years of growth, boom and bust and that important technological innovations seemed to be the beginning of each new cycle. As an example, consider the following cyclical sequence he proposed:

- The first cycle (1780–1842) was triggered in Britain by the first coal-fired smelting of iron ore and the automation of the textile industry, both of which were pioneered in the United States.
- Second cycle (1842–1897) – the age of steam, railroads, and Bessemer steel; this was also a cycle that began in the United Kingdom.
- The third cycle (1898–1939) was marked by the invention of electricity, the development of chemicals, and the invention of motor vehicles. When Kondratieff published his article, a cycle had already begun in the United States and was well underway.

Since then, economists have identified a fourth cycle that spans the period from the conclusion of World War II to the energy crises of the 1970s and beyond. Once again, the United States had a key role in significant technical advances, such as the creation of air transportation, the development of oil-based industries, and the expansion of the electrical and electronic sectors. It is currently believed that a 5th cycle has already begun, and that the new technology heralding this cycle pertains to disciplines such as microprocessors, genetic engineering, new metals, and alternative energy sources, among other things. Japan and the NICS, it might be said, are now in a position to take the initiative in terms of innovation.

Koppie

A little South African rocky hill. The word has, however, been frequently used in Africa for comparable characteristics. Koppies are often made up of joint-limited, frequently collapsed boulders. They are the product of inselbergs prolonged weather, and hence a degenerated form. In reality, the differentiation between koppies and tors is sometimes challenging.

Figure 35. Image showing Koppie.

Sources: Image by Wikimedia.

K-Value

It is used in central place theory to describe the number of central places of a particular order that are reliant on a central place of the next highest order in the central place hierarchy. Christaller has acknowledged three fundamentally distinct k-value systems, each with a different concept and a differing spatial link between the core grid and the hinterland network. These were: k-value 3 – market principle; k-value; 4 – principle transport; k-word 7 – principle administrative.

L

Labor

In the manufacturing process, labor is one of three primary components, the other two being money and land. In order for any production system to function properly, it must be maintained. Labor requirements vary depending on the type of the economic activity involved, with many businesses having extremely particular requirements, such as highly qualified technicians, administrative staff, unskilled manual laborers, or machine operatives, among other things.

A firm's location may be significantly influenced by the spatial availability of particular types of labor skills, as well as by other characteristics of a region's labor force such as pay expectations and dependability, as well as flexibility and adaptability, among other things. It is occasionally the case that the accumulation over time of certain labor skills in a given location can make a substantial contribution to industrial inertia in that region.

Figure 36. Image of Labor working in factories.

Sources: Image by pixnio.

Labor Costs

The business total labor costs, which include pay and salaries paid to its employees, as well as other labor-related expenses like contributions to national insurance, pension schemes, etc. Despite the widespread use of mechanization, assembly-line production, and automation, the expenses of labor continue to constitute a major portion of overall manufacturing costs. This might be a result of the increased demand for better wages and salaries that organized labor has brought to bear in MEDCS, mostly through trade unions, in recent years.

Employee wages and compensation in the United Kingdom now are comparable to 54% of the value created in manufacturing; in the United States, the number is 30% of value added in manufacturing. However, while wage and compensation levels still differ from one activity to the next, it should be emphasized that, as a result of trade union action and the adoption of nationally agreed rates of pay, there is currently very little geographical variation in labor expenses within a single industry as a whole. The fact that there are still salary disparities between men and women exists despite significant progress toward sex equality; this fact continues to convince some businesses to prefer to hire female employees.

Even on a global scale, there are still low-wage places to be found, and as a result, they are appealing to labor-intensive businesses and the TNCS as a result of their cheap wages. The relocation of electrical and electronic companies from the United States and Western Europe to NICS and then to LEDCS during the last quarter of the twentieth century is partly a reflection of the persistent and significant pull of the labor-cost factor on industrial location during the last quarter of the twentieth century.

Laccolith

A kind of igneous intrusion created by the injection of magma along a bedding plane and the resulting doming of the underlying strata, which is concordant with the rest of the world. Some laccoliths have been partially or completely exposed as a result of the removal of weak sedimentary strata, resulting in the formation of enormous domed landforms. Laccoliths of the cedar-tree kind are created when magma, ascending along a central pipe, is intruded into numerous bedding planes inside the up domed structure.

Lacustrine

A lacustrine adjective denoting an association with a present or former lake, and used primarily to denote sediments, such as a lacustrine delta formed by the accumulation of sands and gravels at the point where a stream enters the lake, a lacustrine terrace formed by the formation of beach deposits along the margins of a former lake, and a lacustrine plain, in which the lake has been filled by in-washed sediments.

Lag Industry

An industry that depends on the expansion of one other industry or another sector of the economy for its growth. The production and extraction of sand and gravel from motor vehicle accessories might be mentioned as examples.

Lagging Region

It is possible that a type of depressed region can be distinguished by the fact that, despite the fact that it is performing below the national average in terms of economic growth, it is still making some progress. In that regard the difference may be drawn between a decreasing region in which the fundamental problem is more of an absolute decrease in productivity and wealth; or else there is an ever wider divide between the performance of the region and that of the country as a whole. A lagging area can therefore be considered as less of an area of trouble. In England, regions are falling behind with the status of intermediate areas, whilst the designation of development areas has been reserved for the decreasing areas' even more significant difficulties.

Figure 37. Image showing Lahar.

Sources: Image by Wikimedia.

Lahar

A fast flow of mud and volcanic material caused by the overflow of a crater lake or lava-dammed lake, or the saturation of the sediment (e.g., volcanic ash) by continuous heavy rain. When it is heated, the lahar's velocity can reach speeds of up to 100 kilometers per hour, and the flow can extend for 100 kilometers or more. Lahars frequently engulf farms and communities, resulting in the death of dozens of people in highly populated areas such as Indonesia's Java. Armero, Colombia, was completely destroyed by a catastrophic lahar in 1985, precipitated by the eruption of Nevado del Ruiz and burying portions of the town in up to 8 meters of mud. An estimated 22,000 people died, and the town was completely submerged in mud.

Lame Duck

The type of economic activity that is unable to compete on the free market without the help of the government is known as Lame duck. In the absence of it, the activity would go out of business, resulting in the loss of employment and income. The question of whether or not governments should provide assistance to lame ducks is a contentious one.

The government assistance, some would say, should be directed on stimulating development rather than assisting failing businesses or industries. Others, on the other hand, believe that the loss of a single activity has the potential to trigger a much wider and more dangerous downward spiral of deterioration. It follows that the government's resources will be stretched even more, as seen by the increase in demand for unemployment and other social benefits, among other things.

Laminar Flow

A kind of flow for example, in a river in which water is conveyed at low speeds in a smooth straight Channel, with parallel 'layers' of water shearing over one another in such a way that velocity is least near the bed and largest near the water surface. For the vast majority of actual streams, Turbulent Flow takes the role of Laminar Flow, particularly in areas where the bedrock is rocky, and velocities are high.

Land Breeze

An evening breeze moving from the land towards the sea that is pleasant and mild. In contrast to the reduced air pressure above the still-warm water, the land's temperature is slightly boosted by nighttime radiation, resulting in a small increase in atmospheric pressure. As a result, there is a compensating flow of air to the south. Land breezes are often associated with tropical locations, although they may also occur on a warm summer day in the United Kingdom when anticyclonic conditions are calm.

Figure 38. Image showing land breeze.

Sources: Image by flickr.

Landfill Site

A natural or man-made hole in the earth, for example, mines, quarries and gravel pits, which are used for trash disposal. The technique may pose a threat to the environment due to the possibility for contamination. Consider the possibility of polluting groundwater sources, for example. The breakdown of garbage in an anaerobic environment creates methane, which increases the danger of explosions as it decomposes.

Land Reform

In a given area, an alteration to the land tenure arrangements that are currently in place; for example, the dismantling of large estates and the redistribution of the land by sale to existing owner-occupier farmers and former tenants, or the abolishment of the custom of equal inheritance (gavelkind), which has been the source of severe farm fragmentation for centuries. When it comes to land reform in certain former colonial countries, governments have had a greater impact on the process by displacing big landholders or taking over land they had previously owned by expatriates.

Land Tenure

The complex set of land-tenure and usage rights that is especially relevant in regard to agriculture and also housing. Land tenure may take on a variety of various shapes and forms. Among the most common are:

- owner-occupation, in which the user or occupier is also the property's owner.
- tenancy, in which the user or occupier is not also the property's owner and where the user either pays the property's owner rent or makes payment in the form of labor or share cropping.
- use right, in which ownership is not significant and in which a group or individual establishes rights by use (as in SH Common law, absentee ownership, and governmental ownership are all forms of property tenure that may exist alongside each other.

Land Use

Using land for human activity, which is not usually for financial benefit or profit, is referred to as land use. A fundamental difference may be made between land use in rural areas (agriculture, forestry, recreation, and so on) and urban areas (industry, commerce, housing, and so on).

Land-Value Surface

Property prices and rents fluctuate widely from one location to another as a result of geographical differences in the amount of demand and the quality of the land. This results in an uneven surface that is difficult to navigate.

Landscape

It is a phrase that is used to define the sum total of an area's "aesthetic attractiveness." The physical landscape refers to the combined effect of landforms, natural vegetation, soils, rivers and lakes, whereas the human landscape refers to all of the modifications made by people, such as cultivated vegetation, field patterns, communication networks, villages, open-cast mines and quarries, among other things.

Figure 39. Image showing landscape.

Sources: Image by Wikimedia.

Landscape Evaluation

A quantitative approach for measuring the quality of a given landscape as a tool for planning, development, and conservation. Developing standards for evaluation that are not totally subjective is a challenging challenge to overcome though one method is for a number of observers, acting independently, to classify views from selected positions in terms such as unsightly, undistinguished, pleasant, distinguished, superb and spectacular. These value judgments are given numerical values ranging from 0 to 32, and average scores for certain landscapes are calculated.

Landslide

This is a form of mass movement in which the material being displaced keeps its coherence as a single body while moving along a clearly defined sliding plane. Landslides are frequently triggered by significant accumulations of soil water caused by rainfall, springs, or melting snow, among other things. This raises the weight of the sliding mass and, as pore water pressure increases, lowers the friction between the constituent particles of the sliding mass. In particular, when the sliding mass is composed of worn materials sitting on a substratum of clay, the latter property becomes extremely essential. Landslides have the potential to inflict substantial damage to lines of communication such as trains, roads, and other infrastructure. For example, if low-income shanty town housing has been erected on marginal terrain that has steep deforested slopes, it might result in the death or injury of those living in the area.

Lapse-Rate

A measure of how quickly temperatures vary as a function of height, whether in a steady column of air or a rising pocket of air.

Latent Heat

In calories per gramme, heat energy is defined as the heat energy used in changing the condition or phase of a body without altering the temperature of the body. As a result, the latent heat of evaporation is the amount of energy necessary to turn water into water vapor; this energy is then released again if the water vapor is condensed into water droplets or the latent heat of condensation. When latent heat is released in significant quantities, it plays a vital role in sustaining or speeding up growing currents of air in the Earth's atmosphere.

Lateral Erosion

In the creation of meanders or braided channels, erosion is done by a stream that is experiencing lateral or sideways movement. Active undercutting of the stream bank occurs most frequently at the point where a thread of high-velocity flow or the thalweg impinges on the bank. Sometimes lateral erosion will affect a valley-side slope, and if left unchecked for a long period of time, it will result in the formation of an extensive surface of lateral planation, which will be topped by steep-sided residual hills or remnants of the former interfluves, as shown in the illustration. In the evolution of desert landscape, it is commonly thought that lateral planation, caused by sheet floods and stream floods, may have played a significant role.

Lateral Moraine

It is a clump of unsorted granular debris that has collected around the edges of a valley glacier. It is most frequently formed by weathering and collapse of the slope above the ice and is derived from this. A lateral dump moraine may build between the glacier and the valley wall if the debris is in transit on the ice, or it may collect as part of a supraglacial lateral moraine while the glacier is in transit.

Laterite

A word that may be used in two different ways. In the first instance, it is used to denote tropical soils that have been lateralized; in the second, it is used to describe a hardened layer of ironstone that has formed as a result of the precipitation of iron minerals, either at the surface or in the subsoil, and which has resulted in a concrete-like layer or series of massive concretions.

Figure 40. Image showing laterite.

Sources: Image by Wikimedia.

Laurasia

The ancient northern supercontinent that included North America and Greenland, Europe, and most of Asia, and was divided from the southern continent of Gondwanaland by the old Tethys Sea. During and after the Cretaceous epoch, Laurasia was divided by the westward drift of North America and the opening up of the North Atlantic, both of which occurred simultaneously.

Leaching

The removal of dissolved chemicals, notably bases, by organic solutions known as leachates, which are frequently produced from the decomposition of surface litter. Well-drained, sandy soils and a climate with more precipitation than evaporation are the conditions most frequently linked with leaching. Leaching results in an eluviated horizon that gets progressively acidic and, as a result, less productive as the bases are removed from the soil. When it comes to the production of latosols and podzols, leaching is an essential step.

Leisure

A period of time without work. With the introduction of shorter working hours and early retirement in the medical field, leisure time is rising, allowing for increased involvement in a wider range of viable recreational activities. At the same time, however, this increased leisure is producing a significantly increased demand for recreational amenities such as football fields, tennis courts, golf courses, country parks, swimming pools, and marinas, among other things. For geographers, aspects of leisure that are particularly important include the accessibility of leisure amenities and the degree of agreement between the distribution of leisure demand and the distribution of leisure amenities.

Levee

An enclosing bank of sediment developed along the margin of a river channel occupying a flood plain. Large amounts of silt are deposited on the plain when a flood occurs, and the river overtops its banks and inundates the area. The largest volumes of deposition, on the other hand, occur in a concentrated zone at the boundary between the fast-flowing water in the channel and the slacker water over the flood plain. As a result, the levees gradually rise in height until they reach a level that is higher than the flood plain. If the levees are breached or topped, the potential size of the flood is widened even more. Natural levees are frequently reinforced and artificially elevated in order to decrease the possibility of floods occurring.

Figure 41. Image showing levee.

Sources: Image by Wikimedia.

Lichenometry

A technique of dating rock surfaces that relies on the size and presence of lichens growing on the rock as indications of the age of the rock. When grown in an unchanged environment, lichens (such as the commonly used *Rhizocarpon geographicum* survive for hundreds of years and are the principal colonisers of rock surfaces. Lichens develop slowly but consistently in an unchanging environment.

As a result, the size of a lichen, and particularly its diameter, may be used to determine its age. In lichenometry, growth curves are constructed by measuring lichens on surfaces that are known to be old such as buildings and tombstones; on the Graph, diameter is plotted versus time. Lichenometry may be used to determine the age of a surface.

If the age of a site is unknown for example, a glacial moraine or an abraded rock surface exposed by glacial retreat, a number usually five of the largest and therefore oldest lichens presents are measured and averaged, and the age of the site is calculated from the growth curve of the largest and oldest lichens present. During the Postglacial era, lichenometry has been widely utilized to solve the problem of dating the retreat stages of ice sheets and glaciers, but it may also be used to establish the chronology of changes in other landforms, like the river channels.

Life Expectancy

Life expectancy is a forecast produced on the basis of life table calculations about the average number of years a person could expect to live. Improved food and medical advancements have contributed to a considerable increase in life expectancy in most regions of the world, with women generally living longer lives than males. In the United Kingdom, males have a life expectancy of 70 years and women have a life expectancy of 76 years; in Bolivia, these figures are 42 and 48 years, respectively. The influence of infant mortality causes life expectancy to be higher a year after birth than it is at birth; subsequently, expectancy falls with age.

Lifestyle

A comprehensive pattern of day-to-day activities that constitutes a person's way of existence Age, clothing, place of residence, kind of employment, socio-economic position, leisure activities, and the amount of money spent on products and services all reflect this trend.

Light Industry

The production of items with a tiny volume and a small number of raw materials, such as tools and TVs, as well as the preparation of food, is an example of small-scale manufacturing.

Limestone Pavement

A bar calcareous surface, level and slightly sloped by several small-scale solutions. The placement of limestone pavements appears to correspond with the exposure of large bedding planes as a result of glacial erosion or glacial stripping. When applied to the lee slopes of limestone hills that have been overwhelmed by ice sheets, the technique has proven to be particularly successful.

Using the rectilinear jointing pattern, solutional processes are able to construct blocks known as clints, which are divided by larger joints known as grikes. Eventually, the destruction of the clints and the expansion of the grikes caused by continued solution will result in the formation of a debris layer on which plants can thrive. In this respect, limestone pavements are rather short-lived landforms, and many of them will be lost in the not-too-distant future.

Figure 42. Image showing limestone pavement.

Sources: Image by Wikimedia.

Limiting Angle

A slope angle that specifies the above or below boundaries that operate specific operations. For instance, scree paths are usually confined to 36–

38°, as instability above this angle leads to rapid waste sliding. In contrast, the active SOIL slide is unusual on paths under 20°C, and in the present-day Arctic settings, solifluction may require a 6–8° slope under periglacial circumstances.

Linkage

In the same line of manufacturing or service the link between firms is engaged. No company is completely autonomous. Therefore, every company has unavoidably relations with other companies, either because they receive products and services or because they give goods and services to other companies. The connections of an enterprise are therefore of two sorts, namely reverse or input connections where the company 'receives and forward or output connections where the company supply.

Functional ties will vary greatly in their extent and complexity, but regardless of the area of economic activity, such ties join companies into different-size production chains. The advantage of connecting chains is that specialization, which in turn leads in cost reduction and higher efficiency, is promoted. The drawback of those chains is that the chain is only as strong as the weakest link; failure of one company is to the detriment of another associated company along the chain.

Listed Building

A building of particular architectural or historical importance and special protection under the British planning system, as a consequence of being 'listed' by the Department of the Environment. The listing really leads to preservation or preservation.

Literacy

The capacity of a person to read and write is referred to as literacy. The rate of adult literacy in a nation is a frequently used indicator of the country's level of development. The rate indicates the level of education received, which in turn reflects the level of wealth and social values in the nation. At the moment, adult literacy rates vary from close to 100% in the majority of MEDCS to fewer than 30% in a couple of African nations. Women account for over two-thirds of nonliterate adults in the third world, a manifestation of overt gender discrimination in the workplace.

M

Macroecology

Macroecology is the study of ecological patterns and processes that occur on a wide scale at a distance from the individual organism. Single communities and ecosystems are rarely studied in detail; instead, ideas that apply more generally, such as metabolic scaling, extinction risk and diversity gradients are explored in more depth and detail.

Macrotidal

In coastal locations when the tidal range is more than 4 m, this term is used. Tidal currents dominate the processes that take place in macrotidal regions, such as the coasts of the British Isles and the Mediterranean Sea.

Magma

Magma is a superheated liquid or semi-liquid rock that exists under the surface of the Earth. The inner core, outer core, mantle, and crust of the Earth have a layered structure that is made up of several layers. Magma makes up a large portion of the planet's mantle. This lava has the potential to push through fissures or holes in the crust, resulting in a volcanic explosion. Lava is the term used to describe magma that flows or erupts onto the Earth's surface.

Magma is composed of a variety of minerals, much like solid rock. A tiny quantity of dissolved gases, such as water vapor, carbon dioxide, and Sulphur, are also present in the water. Magma is kept in a fluid condition by the extreme heat and pressure found under the Earth's crust.

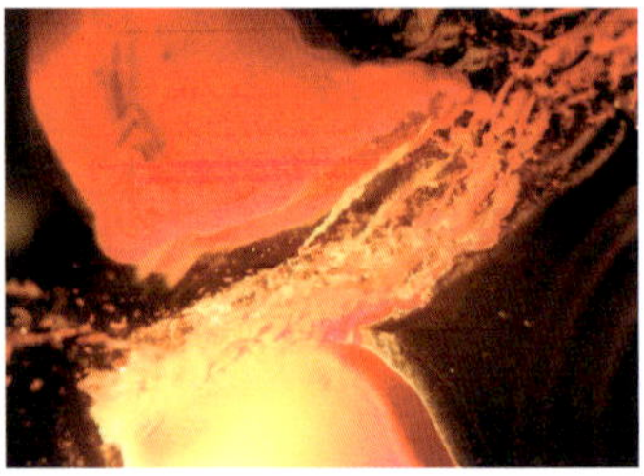

Figure 43. Image showing magma.

Sources: Image by flickr.

Magnetic Polarity Reversal

The change in the polarity of the earth's magnetic field from one polarity to the other is called a magnetic polarity reversal. The occurrence of this has happened at sporadic periods throughout geologic time. In sequences of magnetic rocks, polarity reversals can be retained, which can be compared to normal polarity-change time scales to determine the geologic ages of the rocks. When rocks formed along oceanic spreading ridges cool, they frequently retain this pattern of polarity reversals, and it is possible to use this pattern to estimate the pace at which the ridges are spreading. Sea-floor magnetic lineaments are the reversal patterns that have been documented in the rocks over time.

Magnitude

The magnitude of an earthquake is a numerical value that describes the relative size of the event. The measurement of magnitude is based on the highest amount of motion recorded by a seismograph. There have been several scales established, but the most often used are.

1) Local Magnitude (ML), also known as Richter magnitude,
2) surface-wave magnitude (MS),
3) body-wave magnitude (Mb), and
4) moment magnitude (MM) (Mw).

5)The range and application of the scales 1–3 are restricted, and they are unable to accurately assess the size of the biggest earthquakes. In accordance with the notion of seismic moment, the moment magnitude (Mw) scale may be used to all earthquake sizes, although it is more complex to compute than the other types of scales. In the case of any particular earthquake, all magnitude scales should provide nearly the same result.

Malthusianism

Malthusianism is the belief that population growth has the capacity to be exponential while the increase of food supplies or other resources has the potential to be linear, resulting in a downward spiral of living standards that finally leads to a population extinction.

Mammatus

It is a cellular pattern of pouches hanging beneath the base of a cloud that is generally a cumulonimbus raincloud, but they may be linked to other kinds of parent clouds. Mammatus is derived from the Latin word mamma, which means mammary cloud.

Mangrove Swamps

A coastal wetland prevalent in tropical and subtropical climates, mangrove swamps are a kind of wetlands. The halophytic salt-loving trees, bushes, and other plants that thrive in brackish to saline tidal waters are the distinguishing characteristics of these ecosystems. In estuaries, where fresh water meets salt water, these wetlands are known for their impenetrable maze of woody vegetation, which makes them difficult to penetrate.

They may be found across North America, from the southern point of Florida along the Gulf Coast to the southern tip of Texas. The southwest coast of Florida is home to one of the world's largest mangrove swamps, which is located in the Everglades. Because of their capacity to live in both salt and fresh water, mangrove trees are the dominant species in this wetland environment.

There are just three kinds of mangrove that grow in the continental United States: the red, black, and white mangroves. With its characteristic arching roots, the Red Mangrove (*Rhizophera mangle*) is easily distinguished from other types of mangroves. The black mangrove (*Avicennia* sp.), which is commonly found growing further inland, contains root extensions known as pneumatophores, which aid in the plant's ability to obtain air in submerged soil. White Mangroves (*Laguncularia racemosa*) are able to grow even further inland since their root systems are not very impressive.

Figure 44. Image showing Mangrove swamps.

Sources: Image by Wikimedia.

Manor

The land holdings of a lord, including his demesne and properties within which he has the authority to exercise certain rights, collect certain fees, and so on; historically, the geographical unit of a feudal lordship.

Mantle

The mantle is the mostly-solid portion of the Earth's interior that is responsible for the majority of its weight. The mantle is the layer of the Earth's interior that sits between the thick, superheated core and the thin outer layer known as the crust. There are approximately 2,900 kilometers (1,802 miles) of thickness in the mantle, and it accounts for an astounding 84% of the Earth's entire volume.

At around 4.5 billion years ago, when the Earth began to take shape, iron and nickel separated swiftly from the surrounding rocks and minerals, resulting in the formation of the planet's core. The early mantle was the molten material that surrounds the core during the formation of the planet. The mantle cooled over a period of millions of years. Outgassing is the term used to describe the process by which water contained inside minerals erupted with lava. As more water was expelled into the atmosphere, the mantle solidified.

Manufacturing Industry

Making goods from raw materials via the use of manual labor or equipment is referred to as manufacturing. Manufacturing is often carried out in a systematic manner with the division of labor. Manufacturing is also referred to as manufacturing industry. Manufacturing, in a more restricted meaning, refers to the fabrication or assembly of components into completed goods on a large scale, as opposed to the production of components. The manufacturing industries that create aeroplanes, vehicles, chemicals, textiles, computers, consumer electronics, electrical equipment, furniture, heavy machinery, refined petroleum products, ships, steel, and tools and dies are among the most significant in the world.

Map

An illustration of the earth's surface, or a portion of it, that shows the form and position of different nations, political borders, natural features like as rivers and mountains, and manmade features such as highways and structures.

Map Generalization

The term map generalization refers to all of the changes that occur in a map when a smaller-scale map is derived from a larger-scale map or map data, or vice versa. It is an essential component in cartographic design.

Map Projection

A map projection is a technique for flattening the surface of a globe into a plane in order to create a map. A systematic translation of the latitudes and longitudes of places on the surface of the globe into coordinates on a plane is required for this to be accomplished.

All projections of a sphere onto a plane must, by definition, deform the surface in some way and to some amount, otherwise they are invalid. Because different map projections are used to preserve different properties of the sphere-like body depending on the purpose of the map, some distortions are acceptable and others are not; as a result, different map projections are available to preserve different properties of the sphere-like body at the expense of others.

The characterization of distortions is accomplished through the analysis of map projections. There is no limit to the number of different map projections that can be used. Diverse pure mathematical areas, such as differential geometry, projective geometry, and manifolds, all deal with projections in some form or another. Map projection, on the other hand, refers to a certain type of cartographic projection.

Though its literal meaning suggests otherwise, projection is not restricted to perspective projections, such as those created by casting a shadow on the surface of a screen or the rectilinear picture formed by a pinhole camera on a flat film plate. An alternative definition for a projection is any mathematical function that converts coordinates from the curve to the plane in a discrete and smooth manner. Perspective predictions are a rare breed in today's world of practical projections.

Margin of Cultivation

The distance between a product and a market where the income from a product perfectly equals the entire cost of the product. If production costs are the same no matter how far away from the market a farm is from the market, transportation costs are what define the margin of agriculture.

Marine Erosion

It is the loss or displacement of land, or the long-term removal of silt and rocks along a coastline, caused by the action of waves, currents, tides, wind-driven water, icebergs, or other storm-related effects.

Figure 45. Image showing marine erosion.

Sources: Image by pxhere.

Maritime Transport

In the context of maritime transport also known as ocean transport and fluvial transport, or more broadly aquatic transport, the transportation of people or passengers or products or cargo across waterways is defined as follows- Throughout recorded history, shipping has been a popular mode of transportation for freight. Since the invention of flight, sea travel has lost some of its significance for passengers, but it is still popular for short excursions and pleasure cruises.

Despite fluctuating exchange rates and a surcharge added on top of freighting expenses for carrier businesses known as the currency adjustment factor, shipping by water is less expensive than shipping by air. According to the United Nations Conference on Trade and Development (UNCTAD), maritime transport will account for approximately 80% of worldwide trade in 2020.

A boat, ship, sailboat, or barge can be used to carry goods over large distances via boat, ship, sailboat, or barge, across seas and lakes, through canals, or down rivers. Commercial shipping, recreational shipping, and military shipping are all possible uses for the ocean. The importance of substantial inland commerce has diminished in recent years, but the world's major waterways, which include numerous canals, continue to be highly important and are vital elements of the global economy.

Water transport is capable of moving virtually any sort of cargo; but, when a commodity's delivery is time-critical, such as in the case of certain types of perishable food, water transport becomes impracticable. While sea transport is very cost efficient for regular schedulable cargoes such as trans-oceanic shipping of consumer goods – it is particularly cost effective for large loads or bulk cargoes such as coal, coke, ores, and grains. Most historians agree that the Industrial Revolution occurred most effectively in areas where inexpensive water transportation via canals, navigations, and shippers of various kinds operating on natural waterways enabled cost-effective bulk transportation.

Beginning in the 1970s, containerization radically altered the nature of marine transportation. Items packaged in boxes, crates, pallets, and barrels are included in the category of "general cargo." When a shipment is transported by more than one mode, it is referred to as intermodal or comodal.

Marxist Geography

Using Marxist theories and philosophy to investigate the spatial linkages of human geography, Marxist geography is a subset of critical geography that is gaining popularity today.

Mass Movements

Mass movements also known as mass-wasting are the down-slope movement of regolith i.e., a loose uncemented mixture of soil and rock particles that covers the Earth's surface caused solely by the force of gravity and without the assistance of a transporting medium such as water, ice, or wind also known as mass-wasting.

Muscle Mass

Muscle mass refers to the physical size of a muscle; muscles can get huge as a result of exercise and intensive physical training, although this is not always the case. Body weight muscle strength is considered to be one of the acknowledged components of overall fitness, which also includes other characteristics such as endurance, flexibility, power, and speed.

Magnetic Polarity Reversal

A field of geography that studies the shape and movements of the earth, as well as its seasons and tides, as well as its measurement and depiction on maps and charts using various projection methods.

Maximum Sustainable Yield

The maximum sustainable yield (also known as MSY) is the greatest amount of catch that may be obtained from a fish or other population over an extended period of time. Given that the word was established prior to WWII, one might argue that fisheries scientists were thinking about sustainability long before it became trendy, but they did not have sustainability in mind when they were thinking about fisheries sustainability.

Meander

It is any one of several regular sinuous curves, bends, loops, twists, or windings that are seen in the channel of a river or other watercourse, and they are known as meanders. It is also used to describe the act of walking slowly. In the course of flowing through its floodplain or shifting its channel inside a valley, a stream or river will swing from side to side, causing it to generate this effect.

Figure 46. Image showing meander.

Sources: Image by Pixabay.

Means of Production

In economics and sociology, the term means of production refers to the physical and non-financial inputs that are utilized in the creation of products and services that have a monetary or other economic value. Raw materials, manufacturing facilities, machinery, and tools utilized in the creation of goods and services are examples of such items.

Mechanical Erosion

Gravity, aeolian processes i.e., those produced by the wind ice in the form of glaciers, and water in the form of rivers, waves, turbidity currents, and runoff generated by rainfall are the primary agents of mechanical erosion.

Mechanical Weathering

The mechanical weathering also known as physical weathering is the process through which rock is broken down into smaller fragments. These tiny bits are identical to the larger rock, except that they are smaller. That indicates that the rock has changed in terms of its physical properties without changing in terms of its chemical makeup. The smaller pieces contain the same minerals as the original rock, and they are in exactly the same proportions.

There are a variety of techniques for breaking down large boulders into smaller bits. If you live in a climate that routinely cycles above and below the freezing point, you are probably familiar with ice wedging, which is the primary mechanical weathering process. Ice wedging is effective in a short period of time, breaking apart rocks in regions where temperatures fluctuate between above and below freezing throughout the day and night, as well as between above and below freezing with the seasons.

Mediterranean Climate

Located between approximately 30° and 45° latitude north and south of the Equator and on the western sides of continents, the Mediterranean climate is the most common climate type according to the Koppen classification. It is characterized by hot, dry summers and cool, wet winters, and it is the most common climate type in the world. It is subdivided into the CSA and CSB subtypes according to the Koppen-Geiger-Pohl classification system.

A subtropical anticyclone's extension and expansion poleward across the seas bring sinking air to the region in the summer, which is characterized by clear sky and high temperatures. The anticyclone travels Equatorward in winter, and it is replaced by moving frontal cyclones, which bring with them heavy precipitation and wind chills. Generally speaking, annual temperature ranges are smaller than those found in marine west coast climates, owing to the fact that locations on the western sides of continents are not well positioned to receive the coldest polar air, which forms over land rather than over water, and thus experience less extreme weather.

Mediterranean climates are also drier than humid subtropical climates, with precipitation totals ranging from 35 to 90 cm 14 to 35 inches per year; the lowest amounts occur in interior regions adjacent to semiarid steppe climates. Mediterranean climates are also drier than humid subtropical climates.

Mediterranean Soils

Mediterranean soils are soils that have formed as a result of living in a Mediterranean climate. Terra Rossa on hard limestone and Red Mediterranean Soils are two names for these types of soils. If we look at the land usage and production potential of these soils, it can be seen that they are heavily exploited for both rain-fed and irrigation-fed crops.

Megalopolis

One or more roughly adjacent metropolitan areas, known collectively as a megalopolis; also known collectively as a megaregion; or known collectively as a supercity; are grouped together in a megalopolis. These metropolitan areas may be somewhat separated or may merge to form a continuous urban region.

Figure 47. Image showing megalopolis.

Sources: Image by Pixabay.

Memory

Memory refers to the procedures that are utilized to acquire, store, retain, and retrieve information in order to perform subsequent operations on it. Memory is comprised of three primary processes: encoding, storage, and retrieval. Encoding is the process through which information is stored. The ability to both retain and retrieve knowledge that we have acquired or experienced is a crucial component of human memory.

Mental Mapping

An individual's point-of-view impression of their region of contact is referred to as a mental map in behavioral geography. Despite the fact that this type of subject matter would appear to be most likely to be researched by areas in the social sciences, geographers are the ones who are most frequently involved in this type of research.

Mesa

An isolated, flat-topped landform that rises sharply above the surrounding land area and is bordered by steep slopes or cliffs; it is typically capped by erosion-resistant, almost horizontal rock; and it is found in arid regions often lava. Mesas and buttes are similar in appearance and are found in isolated locations. A mesa has a peak area that is larger than the height of the surrounding cliffs.

Mesas are most frequent in arid and semiarid climates, but they are not confined by climate in any way.

Metamorphic Rock

Rock of any origin that has been changed in mineralogical composition, chemical composition, or structural composition by heat, pressure, and movement deep below the earth's crust is known as metamorphic rock. Almost all of these rocks have a crystalline structure. For example, schist, gneiss, quartzite, slate, and marble are all types of rocks.

Metasediment

A sedimentary rock or a sedimentary deposit that has been exposed to metamorphism and exhibits signs of having done so.

Meta-Stable Slope

When the environment's equilibrium is upset, such as by road building or the removal of vegetation, a slope that is reasonably stable at the present moment may become unstable and become active again. The base levels of previous geomorphic events are frequently associated with a metastable slope. It is common for the regolith to be relatively deep, and it may contain stone lines or relict traces of slope alluvium, among other things. The gradient of a slope is typically between 15 and 45%. Active slope is a good comparison.

Meteorite Crater

Barringer Crater, for example, is an impact crater produced by a huge meteorite striking the earth's surface and causing it to collapse.

Figure 48. Image showing meteorite crater.

Sources: Image by Wikimedia.

Microbiotic Crust

When soil particles are held together largely by living organisms and their organic wastes, the result is a thin, surface layer or crust of soil particles; its thickness can range from 1 cm to 10 cm, and its aerial coverage of the ground surface can range from 10 to 100%. Crusts act as a stabilizing agent for loose earthy material. Chemical crusts for example, salt crusts and physical crusts are two more forms of surface crusts to consider, e.g., raindrop-impact crust.

Microfeature

Localized natural formations (features) on the ground surface that are too tiny to be shown on a topographic or soils map at the scales typically employed for topographic or soils mapping, example, 1:24,000 to 1:10,000. Earth pillars, patterned ground, and frost boils are examples of such structures.

Microhigh

This is a general microrelief word that refers to regions that are slightly raised relative to the nearby ground surface; the variations in relief can range from a few centimeters to several meters in elevation. Cross-sectional profiles can be simple or complicated, but they are often comprised of softly rounded, convex tops with gently sloping sides; this is referred to as micro-high in certain circles and spelt micro-high in others.

Microlow

This is a generic microrelief word that refers to regions that are somewhat lower than the neighboring ground surface like shallow depression; the variations in relief can range from a few centimeters to several meters in elevation. Simple or complicated, cross-sectional profiles are typically comprised of muted, concave, open or closed depressions with gradually sloping sides; this is also referred to as micro-low in some circles.

Microrelief

A land surface's height changes that are too tiny or complex to be shown on a topographic or soils map at widely used map scales are referred to as microtopography (e.g., 1:24,000 through 1:10,000). Micro high, micro slope, and Microlow are all examples of microstructures. Microfeature that allows you to compare two things.

- Compare microfeature SW.
- SW not preferred.

In general, it refers to small, localized abnormalities in the shape and height of a land surface that are overlaid on larger landforms, and it includes characteristics like as low mounds, swales, and shallow pit, among other things.

Micro Slope

A general microrelief phrase that refers to areas of nominal surface relief (slightly sloping to level) when compared to the neighboring ground surface; variations in total local relief can range from a few centimeters to several meters in magnitude and location. It is possible to have simple or complicated cross-sectional profiles, but the most common are low and softly rounded convex tops Microhigh, gently sloping to level sides (micro slope), and depressional low regions i.e., Microlow. Typically, micro slopes account for the vast bulk of the land surface area in gilgai and other environments characterised by microrelief.

An ancient settlement's midden is a mound or strata of rubbish or broken pots, ashes, food remnants, and so on that is typically discovered on the site of the settlement's origin.

Mima Mound

Unstratified, loose, gravelly, silty, or sandy material is utilized to form a variety of low circular or oval domes that may be seen across the world. The diameter of the base can range from 3 meters to more than 30 meters, and the height can range from 30 centimeters to about 2 meters.

Figure 49. Image showing Mima mound.

Sources: Image by Wikimedia.

Mine Spoil

Mine spoil is a term used to describe a mixture of earthy elements that have been intentionally deposited as a result of coal mining activities, whether on the surface or below.

Mine Spoil, Metal-Ore Extraction

Mine spoil is a term used to describe a mixture of earthy elements that have been intentionally deposited as a result of metal-ore mining activities, whether on the surface or below.

Mine Spoil or Earthy Fill

A pile of earthy material, rock, or other waste material that has been displaced during mining or excavation operations.

Minor Scarp

It is a steep slope formed by differential motions inside the sliding mass that forms on top of displaced material after a landslide.

Miocene

This is the geologic era that immediately follows the Oligocene Epoch and precedes the Pliocene Epoch about 5.2 to 23 million years ago. It is also the time-stratigraphic series of earth materials that corresponds to this epoch nearly 5.2 to 23 million years ago.

Mogote

A karst tower is a type of isolated, steep-sided, commonly asymmetrical hill or ridge composed of limestone that is generally steeper on its leeward side or prevailing downwind side and surrounded by a near level to sloping coastal plain composed of marine and alluvial sediments; it is a type of karst tower that is composed of limestone.

They range in height from a few feet or less than one meter to more than 150 feet or 50 m. However, some create clusters of hills or ridges that rise out of the surrounding blanket deposits, while others are solitary and cover only a limited area. Mogotes may be found in large numbers across northern Puerto Rico.

Monadnock

In a temperate environment, an isolated hill or mountain of resistant rock that rises noticeably above the general level of a lower erosion surface is an isolated relic of a past erosion cycle in an area that has been mostly beveled to its base level.

Monocline

Folded strata that dip from the horizontal in just one direction, are not part of anticlines or synclines, and are found at the surface of the earth's crust. As a rule of thumb, this structure can be seen in plateau areas where almost flat strata locally adopt severe dips as a result of differential vertical motions that do not include faulting.

Moraine

Glacial till is an unsorted and unstratified glacial drift that is formed largely by the direct action of glacier ice. It may be found in a variety of landforms, including mounds, ridges, and other topographically different accumulations. An all-encompassing word for a landform that is mostly made of till that has been deposited by glaciers; also known as kame. In external shape, moraine is similar to other forms of moraines; nevertheless, it is comprised mostly of stratified outwash sediments, as opposed to other types of moraines. Moraines are classified into the following types: disintegration, end, ground, kame, lateral, recessional, and terminal.

Moss Peat

It is a collection of organic material that is primarily composed of the remnants of mosses (e.g., sphagnum moss). Herbaceous peat, sedimentary peat, woody peat, mucky peat, and mucky peat are all different types of peat.

Mound

- A low, rounded natural hill of unspecified origin, generally less than 3 m high and, composed of earthy material.
- Or, A small, human-made hill, composed either of debris accumulated during successive occupations of the site or of earth heaped up to mark a burial site or burial mound.
- Or A structure built by colonial organisms like termite mound.

Mountain

When the land surface rises more than 300 meters above surrounding lowlands, it is referred to as a mountain range. Mountain ranges typically have a nominal summit area relative to bounding slopes, and they typically have steep sides greater than 25% slope, with or without significant bare rock exposed. A mountain can be found as a single, solitary mass or as part of a group of mountains that form a chain or range. Mountains are produced largely as a result of tectonic activity and/or volcanic activity, with differential erosion playing a secondary role.

Figure 50. Image showing mountain.

Sources: Image by pixnio.

Mountain Base

Long transport colluvium and slope alluvium sediment make up the lowermost region of a mountain's geomorphic component, which includes the firmly to slightly concave colluvial apron or wedge at the bottom of mountain slopes, as well as the lowermost area of the mountain's lowermost area. It can spread out into more level valley regions, where it eventually interfingers with, is buried by, or is replaced by re-emergent residuum, before disappearing altogether. Mountaintop, mountain flank, free face, geomorphic component are all terms that may be used to compare.

Mountain Flank

Mountains side area is characterised by extremely long, complex backslopes with relatively steep slope gradients, and is composed of highly-diverse, colluvial sediment mantles, complex near-surface hydrology, mass movement processes and features like creep, landslides; rock outcrops or structural benches may also be present. The mountain flank can be split according on where it is located along the mountainside in general i.e., upper third, middle third, or lower third mountain flank. Mountaintop, mountain base, free face, and geomorphic component are all good comparisons.

Mountain Range

Mountains are a single, massive mass composed of a series of mountains or tightly spaced mountain ridges, with or without peaks, that are closely connected in position, direction, orientation, creation, and age; mountains are also considered to be a component portion of a mountain system.

Mountains

The presence of mountains and the valleys that separate them; a generic term for any group, cluster, or sequence of mountains, narrowly spaced mountain ridges, with or without peaks, closely related in position, orientation, direction, formation, or age; and a mountain range whose summits commonly exceed 300 metres in elevation approx. 1000 ft.

Mountain Slope

A section of a mountain that is between the peak and the base.

Mountain System

In geology, a collection of mountain ranges that have certain unifying characteristics, such as a resemblance in shape, structure, and alignment, and which are thought to have originated from the same basic processes is referred to as an orogenic belt.

Mountaintop

A geomorphic component of mountains that consists of the topmost, generally flat or gently sloping region of mountains, which is characterised by relatively short, simple slopes comprised of exposed rock, residuum, or short-transport colluvial deposits. Mountaintop soils may be fairly thick and highly formed in humid climates, making them ideal for agriculture.

Mountain Valley

- If a stream cuts or deepens a tiny, externally drained V-shaped depression in cross-section and floors it with alluvium, or if an alpine glacier modifies a larger, U-shaped depression that floors it with either till or alluvium, it is referred to as an alpine glacial till depression. Mountain valleys may be classified into several categories depending on their morphology and the sediments that make up their valley floors that is V-shaped valley, U-shaped valley.

  Or,

- A relatively small, structural depression inside a mountain range that is partially filled with alluvium and drains externally to an intermontane basin, bolson, or semi-bolson is known as a bolson or semi-bolson.

Muck

A kind of soil material that is largely made of severely degraded organic material and in which the original plant components are no longer discernible, for example, sapric soil materials of Soil Taxonomy. When compared to peat, it often includes more mineral content and fewer plant fibers, and it is typically darker in color. Peat, mucky peat, and herbaceous peat are all different types of peat.

Mucky Peat

Organic matter dominates unconsolidated soil material that is at an intermediate stage of decomposition, so that a large portion of the original material can be identified but a considerable portion of the original material cannot be identified, for example, hemic soil materials of Soil Taxonomy. Peat, muck, and herbaceous peat are all different types of peat.

Mud Flat

It is defined as a generally flat region of fine-grained material like silt along a beach as in a protected estuary or surrounding an island, which is alternately covered and exposed by the tide or covered by shallow water, and which is devoid of any plant growth.

Figure 51. Image showing mud flat.

Sources: Image by Wikimedia.

Mud Pot

A sort of hot spring that contains boiling mud that is typically sulphureous and colourful, such as that found in a paint pot. Mud pots are frequently found in volcanic environments, particularly in Yellowstone National Park, WY, where they are frequently linked with geysers and other hot springs.

Mudflow

It is characterised by a highly fast type of earthflow that is dominated by abrupt, downslope movement of a saturated mass of rock, dirt, and mud with more than half the particles being less than 2 mm in size that moves as if it were a viscous fluid when it is in motion. Compare and contrast – debris flow, flow, and landslide.

Mudstone

A fine-grained sedimentary rock that is blocky or massive in appearance and in which the proportions of clay and silt are roughly equal.

Or,

The word claystone refers to a broad range of clay and silt-containing rocks such as claystone, siltstone, shale, and argillite, and it should be used only when the quantities of clay and silt are unknown or cannot be determined with precision.

Muskeg

A bog, generally a sphagnum bog, with grassy tussocks or hummocks growing in moist, poorly drained northern regions, with deep accumulations of organic material, sometimes in areas of permafrost; a moss-covered muck or peat bog of boreal regions.

N

Natural Levee

As a result of floodwaters overflowing the typical banks, a stream builds a long, broad low ridge or embankment of sand and coarse silt along both sides of its course, particularly in times of flood when it is compelled to deposit the coarsest part of its cargo. It has a moderate slope away from the river and into the adjacent floodplain, with its maximum elevation being closest to the river bank. It is surrounded by a floodplain.

Figure 52. Image showing natural levee.

Sources: Image by Wikimedia.

Nearshore Zone

The term subaqueous marine or lacustrine landform area" refers to a landform region that typically parallels or extends seaward or lakeward from the low water line to beyond the breaker zone, which may include longshore bars. During the journey to the beach, the waves steepen, break, then reform in the nearshore zone of the ocean. Through the action of waves and currents, sediment is transported both parallel and perpendicular to the coast.

Nearshore Zone (Relict)

As a result of isostatic rebound or glacial lake draining, a previous nearshore zone has become subaerially exposed. The previous landward boundary of a relict nearshore zone is often marked by an elevated beach; in some cases, relict longshore bars can be found in offshore locations as well. Surficial sediments may show signs of wave and current activity, such as sorting or particle-size discontinuities, which indicate the presence of wave and current action.

Neck

It means volcanic neck or patterned ground.

Nivation

The procedure of removing the fine particles along the border of the decreasing snow patch or snow bank, especially by washing sheets, by the snow flow and by the solution of the meltwater, for a low depression or nivation hollow on a mountain-side. The action of freezing and thawing appears to be minor.

Nivation Hollow

Slightly or occasionally occupied by a snow bank or showplace or a snow pit on a mountain side formed by nivation and not cliffed depression. If every summer the snow fades entirely, the hollow becomes deepened; otherwise, it cannot be a circus forerunner if the alpine glacier progressively enlarges and deepens it.

Figure 53. Image showing nivation hollow.

Sources: Image by flickr.

NAFTA (North America Free Trade Agreement)

Canada, the United States and Mexico are members. It was founded in 1988 with the aim of developing an EU-wide alternative Single Market in North America. Two of the so-called global trinity are NAFTA, together with the EU the third is E and SE Asia.

Nappe

A huge recumbent fold, broken by a thrust fault at a low angle, with its several kilometers of forwards rocks on the upper limb above the fault.

Figure 54. Image showing nappe.

Sources: Image by Wikimedia.

Nation

A group of individuals, often identified as a distinct entity, related to one another via links between history, feeling, descent and occasionally language. The colonial activity of the imperial powers has also established nations, for example, in Africa. A nation having a separate autonomous state as is the norm at UN is usual practice; nevertheless, a State might really consist of two or more national groupings. For, example, the Walloons and Flemings in Belgium; the Czechs and Slovaks in what was Czechoslovakia.

National Grid

The ordnance survey maps utilise a grid that is based on the transverse Mercator map projection and is called a transverse merchant.

- The Central Electricity Generating Board (CESB) provides an easy and accurate reference system, allowing the precision to define any site by a number and letter code i.e., a combination of the east and north sides read from the map margin.
- The Central Energy Generating Board (CEGB) established a network for delivering electricity to all regions of the United Kingdom in the early 1900s.

National Park

For the conservation of the scenery, vegetation, wildlife, and historic objects an area is set aside by planning the controls and other regulatory devices which will leave them unimpaired for the future generations and along with this also for the public enjoyment and scientific purpose. It has been seen that from country to country the status and specific aims of national parks vary.

In the year 1872 it has been seen that with the designation of Yellowstone the USA the concept began. It has been noted that there are now some 35 parks, mainly wild areas, embracing about 60,000 km2. As by following the National Parks and Access to the Countryside Act (1949) national parks were first established in England and Wales. To minimize the impact of land use is the main aim on the intrinsic character of scenic landscapes.

It has been noted that there are 10 designated parks (including the Lake District, Snowdonia, and the Pembrokeshire coast), covering an area of over 13,000 km2, but two more areas (the Norfolk Broads and the New Forest) have recently been given a status closely akin to that of a national park. For the protection of animal life, the national parks have also been established in the third world countries, as in Kruger (S Africa), Serengeti (Tanzania) and Tsavo (Kenya).

Nationalization

It has been seen that from the private sector to public sector there is transferring of the production of goods and provision of basic services. It has been seen that effort has been made to extend the state ownership of the economy.

NATO

Countries like Belgium, Canada, Denmark, France, Iceland, Italy, Luxembourg, The Netherlands, Norway, Portugal, the UK, and the USA signed the NATO in the year 1949 and created by a defence. In the strategic arena of Europe and the Atlantic the military strength of the Soviet bloc and the perceived threat which is being posed and also the principal aim was to counteract it. With the break-up of the Soviet Union and the decline of communism in the year 1990, but to include some former members of the Soviet bloc., broadening of its membership is being required.

Natural (Environmental) Hazard

To the human settlements, man-made structures, economic activities, etc. There is natural occurrence that threatens, or actually causes damage and destruction. River floods, earthquakes, tsunami, volcanic eruptions, landslide and glacial out bursts are some of the most serious environmental hazards. In the modern geographical study, the ways in which people react to such hazards is an important focus. Due to the human activities, it has been said that it is also worth remembering that there are hazards.

Natural Increase

It has been noted that from an excess of births over deaths the growth in the population of an area has been resulted, which is different from the growth attributable to a positive migration balance.

Natural Resource

It has been said that for the people any commodity that exists in the natural state is useful, e.g., minerals, rocks, SOIL, water, plants, animals and air. As it depends on ability of people to discover the resource that whether it is exploitable or not and also provide the necessary technology to exploit it. The link between the people and physical environment is being provided by the natural resources and also viewed collectively.

The difference between renewable and non-renewable resources which is being made is also important. By rising levels of population and technology the former (e.g., water and air), although inexhaustible, are subject to abuse (e.g., pollution), while the latter (e.g., minerals and soils) are finite, exhaustible and being depleted at increasing rates.

Natural Vegetation

By the humans and their activities vegetation has been unaffected. In the perfect equilibrium it seems likely that true natural vegetation with the existing climatic and edaphic conditions, is rare or non-existent, certainly in areas such as Europe that have been occupied and utilized by humans for thousands of years. For this good example is that what appears natural vegetation and which is much altered.

Figure 55. Image showing natural vegetation.

Source: Image by pxhere.com.

Nature Reserve

There is an area which is preserved that botanical and zoological communities may survive. This does not mean an untouched wilderness, but to maintain a particular habitat the organized protection of rare or endangered species may involve the careful control. To limit public access to nature reserves and so to minimize disturbance regulations are strictly imposed naturally.

For the nation and for a network of National Nature Reserves Wales are responsible in Britain, English Nature, Scottish Natural Heritage, and the Countryside Council and on the other hand Royal Society for the Protection of Birds (RSPB), the largest voluntary conservation organization in Europe, holds around 150 reserves.

Neap Tide

In the quadrature when the Earth, the Sun and the Moon are there and tide generating forces are at minimum. It has been said that high tides will be relatively low and low tides will be relatively high and in other words the tidal range will be much reduced. With the first and last quarters of the moon neap tides occur every 14.75 days.

Nearest-Center Hypothesis

To minimize the distance travelled the essence of this hypothesis is that consumers of goods and services seek in order to secure a particular good or service. In other words, it can be said as that goods and services will be obtained from the nearest outlet in order to keep. So, there is an assumption that, by the results of studies which show that consumers do not always behave in this economically rational way has been seriously challenged. Relatively, the consumers' perceptions and preferences are being reflected by the patterns of the consumer behaviour and the personal mobility and affluence, etc. So, from these studies what does emerge is that awareness of the desirability of reducing distance travelled is strongest where low order goods and services are being sought.

Neck (Volcanic Neck)

The pipe of a former active volcano is being occupied by a plug solidified lava and to removal of the surrounding material by denudational processes frequently exposed as a striking landform owing.

Need-Satisfaction Curve

Between the level of living and wellbeing this graph shows the relationship, with the former being seen as an input to the latter. It has been said that four different states of wellbeing are defined along with the y axis. To sustain a given output or state of well-being the curve shows the inputs or level living. As the level of living increases the form of the curve indicates that the impact on well-being is less than proportional. It has been seen that there is the equal spacing between A and B, B and C along the y axis is not repeated along the x axis. Once the hunger is being satisfied the essential message is being conveyed and to improve the wellbeing there is further increase in food supply which are unlikely to do much. Certainly, a point will come when the excessive feeding has detrimental impact in health.

Needle Ice (Also Known As Pipkrake)

When freezing affects the topmost layer of the soil then a small ice crystal up to several cm in length is formed. Above the surrounding soil surface, the needle ice is often seen raising individual stones or soil particles. In the successive freeze-thaw cycles the development and melting of needle ice in is an important factor in solifluction creep in cold climates.

Negative Feedback

It is termed as mechanism by which the open system undergoes the self-regulation. In simple terms, as a feedback loop negative feedback can be defined that counter or works against the processes operating, such that their effectiveness is reduced. For instance, with steep beach gradients on a beach, destructive ways tend to be associated.

Therefore, in the gradient as material is removed and carried away their action leads to a reduction. It has been said that in this way they become less effective. To move towards an equilibrium, state the concept of negative feedback is important in that it demonstrates the ability of natural systems. Though, this natural form of self-regulation is being disrupted by the people.

Negative Skew

Towards the high values a bias within the distribution, as opposed to positive skew, which is a bias towards low values.

Neighborhood Effect

It has been seen that spatial diffusion process has a key aspect, whereby the adoption of some innovation is likely to occur most readily around those who have already adopted it. Knock on effect or ripple is being created by the adopters.

Neighborhood Unit

In Britain and America, the idea that has considerably influenced the design of post-war residential areas and new towns. In the garden city concept the idea had its roots and experimented with as part of the Radburn layout. Within the towns and cities, the intention is to produce a cellular structure of small-scale communities. By the provision of centralized, low order services and by clearly defining the physical limits of the community the sense of community and focus is being encouraged. In the setting aside of parks and play areas there is concern for the environmental quality is being reflected and through traffic designing traffic flows in such a way that there is no penetration of residential areas.

Neocolonialism

By the so-called superpowers the economic and political control which is exerted over apparently independent LEDCS. It has been said that the control may take a variety of forms, both direct and indirect, from technical to financial aid, from trade to military defence. In supposedly uncommitted areas of the world the ultimate objective is to secure spheres of influence, for example, during the second half of the 20th century as part of the cold war conducted between East and West. It has been noted that activities of TNCS represent a form of neo colonialism by which it is claimed.

Neoglacial

With several minor glacial advances, a period of relatively cold climate is being associated which began approximately 5000–4000BP. As the little Ice age these advances have also been referred, though for the advances that affected many Alpine glaciers between 1550 and 1850 in Europe the term has been reserved. There was a relatively warm period prior to the Neoglacial period known as the climatic optimum.

Neolithic Period

From the latter part of the 4th millennium BC a cultural period following the Mesolithic period until the onset of the Bronze age. To the earlier percussion and pressure-flaking methods it is characterized by the addition of polishing and grinding of stone tools. Its distinguishing feature was the beginning of the domestication of animals, the cultivation of crops and the making of pottery, and in Britain the construction of long barrows, megalithic tombs, and great religious sanctuaries such as Wood henge and the first part of Stonehenge.

Figure 56. Portraying neolithic period.

Source: Image by Designump.

Neo-Malthusians

Towards the end of 18th century Malthus was writing and there are many today who believe that unsustainable rates of population growth are increasingly the cause of much human misery. In the underdeveloped countries each year food production countries falls a bit further behind burgeoning population growth, and people go to bed a little bit hungrier. With the subsistence demands of an increasing population the food supply chains in many LEDC are unable to keep pace.

Neotechnic Era

In Britain the second phase of the Industrial Revolution has started in the year 1870 and also involved in technological advances, such as the increasing use of electricity and the development of the motor vehicle; the proliferation of light, market-oriented industries, the expansion of markets and the rapid outward spread of the built-up area.

Nested Sampling

Into a hierarchy of sampling units, a sampling method that involves dividing a study area that nest within one another. Within these the random processes select the large 1st order units, the middle or 2nd order units and so on. The final location of points within each small, or 3rd-order, unit can be randomly determined. There is a hierarchic sampling or nested which offers the advantage of reduced field costs and of being able to investigate scale variations, but it suffers from high sampling error.

Nesting

In cellular manner there is an arrangement of the market areas or hinterlands in which those of a low order fit, within the higher-order units. For example, the hinterland of, say, a city will comprise the nesting hinterlands of its dependent places as per the central place theory.

Network

It is termed as the system of incorporated places, people, or the objects. By the spatial arrangement of nodes, the character of any network is dependent not only on the pattern generated, but also on its density and the connectivity. In terms of length per unit area the density of a network is simply the total length of the network expressed, as it has been seen that in the case of transport network it gives an indication of the degree of accessibility which is also afforded in the area served by the network. Within the network the connectivity gives some measure of the directness of movement that is possible.

Network Analysis

Using a range of possible measures to assess such properties as its connectivity and its accessibility there is the establishment of the fundamental character network over the time and space there are such measures are useful in comparing networks, while topological maps are useful in depicting networks.

New Commonwealth Countries

In the British a difference made in census under the heading of country of birth. The title embraces all Commonwealth countries except the Old Commonwealth countries of Australia, Canada, and New Zealand, i.e., principally Commonwealth countries in Africa, the Caribbean, the Pacific, and S and SE Asia.

New Town

In a relatively short time, a comprehensively planned urban community which is built from scratch and created for some specific purpose, e.g., for a major city providing a residential dormitory and relocating a national set of government. A comprehensively planned URBAN community, built from scratch in a relatively short time, and created for some exploiting the resources of a peripheral region.In the early 1900s the modern new town movement began in Britain with the building of garden cities. There are some 28 new towns were officially designated and built in Britain post-war period and following the New Towns Act (1946), and there are three principal reasons, i) to take overspill from London and other conurbations, e.g., The ring of eight self-contained new towns located around London (Crawley, Harlow, Stevenage, etc.); (ii) to act as a growth pole or catalyst in the planned revival of depressed regions, e.g., Aycliffe, Peterlee, and Washington in NE England; (iii) to accompany a large-scale expansion in basic industry, e.g., Corby and the iron and steel industry in Northamptonshire.

In the development of British new towns which emerged during the 1960s, namely a move to much larger schemes (population targets set in the 150,000–250,000 range rather than around 80,000, as was the case with the first-generation new towns designated during the late 1940s) and the expansion of sizeable existing towns as 'new town' schemes (e.g., as in the cases of Northampton and Peterborough). With a population of nearly 200,000 is probably one of the most successful of Britain's later new towns says by Milton Keynes. Therefore, no new designations have been made since the 1970s.

NGO (Non-Government Organization)

With the significant influence and responsibilities an organization is created by volunteers rather than government legislation. At an international level and helping NGOs are leading charities operating, for example, in major disaster situations (e.g., Oxfam, the Red Cross) or protecting the environment (e.g., Greenpeace, WWF). Others, such as the National Trust and the Royal Society for the Protection of Birds (RSPB) operate at a national level.

NIC (Newly Industrializing Country)

Over the last three decades there is a term used at the time of referring to a select band of developing countries and also have sustained high rates of economic growth. They have put-performed the MEDC, principally as a result of their competitive edge in manufacturing. In E and SE Asia – Hong Kong, Malaysia, Singapore, S Korea, and Taiwan – sometimes referred to as the Asian Tigers there is cluster of NICS.

Nimbus

It is considered as a type of cloud from which the rain is falling. With the other cloud types the term is usually used in conjunction, as in cumulonimbus and nimbostratus and usually in association with the warm front a uniform, low, dark-grey cloud, giving continuous rain or snow.

NIMBY

Short for 'not in the back yard.' On the grounds it has been seen that a term is being used where the object to a particular development is too near to where they live and will disturb them. The classic example is the bypass proposal where there is general agreement about the need to build one, but a precise routing cannot be agreed because everyone objects to it running past their property.

Nitrates

From the use of nitrogen chemicals are derived as a fertilizer in farming, sinch in the rivers and aquifers the advent of intensive farming has become increasingly concentrated. For the process of eutrophication of rivers and lakes nitrates are thought to be responsible. On the human health some people believe that there may be harmful effects, as in water with links being suggested between nitrate concentration and the incidence of stomach cancer and the 'blue-baby syndrome.'

On the use of fertilizers and reducing the intensity of farming it is now generally agreed that levels of nitrates should be reduced by cutting down. Therefore, through the ground to groundwater it takes up to 40 years, as percolation of water containing chemicals, such as nitrates and in underground water supplies it will be some time before the effects of any change will register.

Nitrogen Cycle

With the nitrates a circulation of nitrogen that starts in the soil being taken up by the roots of plants. It has been seen that by animals and the nitrogen released as ammonia in their excretions these plants are then consumed. On the soil and bacteria, the excreta.

A circulation of nitrogen that starts with nitrates in the soil being. These plants are convert the ammonia back into nitrates, which are then leached back into the soil. And after this cycle then starts then again.

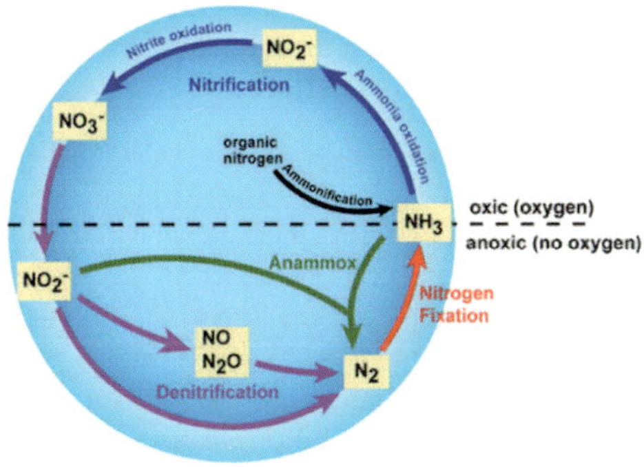

Figure 57. Nitrogen cycle representation.

Source: Image by flickr.com.

Nivation

As the snow patch erosion sometimes, it is referred as complex group processes involving the freeze-thaw action, solifluction, transport and by running water and chemical weathering beneath and at the edges of the snow patch which is being formed in a hillside hollow. To the frost action seasonal enlargement and shrinkage of the snow patch, surrounding areas will be exposed and the resultant debris will be removed (mainly in the thaw season) by solifluction and rivulets of meltwater escaping from the toe of the snow. It has been seen that into the hillside the snow patch will eat back over the long period forming a rounded nivation hollow with a comparatively steep head. In the formation of the nivation cirques and the true glacial cirques, they are being regarded as early stages.

Nivation Cirque

It is referred as a transitional form between a Nivation hollow and a glacial cirque. Some times more than 100 m in height a Nivation cirque possesses steep, frost-shattered headwalls, and in plan are semi-circular. In the northern hemisphere most showed a preferred orientation, facing the north-eastwards. Along with this Nivation hollows do not possess rock basins and lips in contrast to fully developed. They are, therefore, the surface of the snow patch to accumulate along a line roughly parallel to the headwall.

Niveo-Fluvial

To describe a group of periglacial processes a term is being used that is operated in England at the close of the final glacial period. With heavy winter snowfall on high ground there is a relatively mild climate but a rapid early summer thaw, resulted in many fluctuations of temperature about 0°C, intense frost weathering, rapid solifluction, and periods of powerful surface run-off. At some points this led to heavy and selective erosion.

Nodality

Any set of things having a linear character approach the degree or extent to which lines, roads, or any each other or converge at a point.

Nomadism

By people a way of life followed and animals, and involving more or less perpetual movement in search of food and in response to various pressures.

Nominal Data

It is one of three major categories data. To the classification it includes data referring that do not imply quality or relative order of magnitude. Between the forest, grassland, and heathland ecosystems there is difference or between settlements of different shapes which would constitute the nominal data. Frequency is the most common statistical measure of the nominal data which means the number of times a particular event occurs.

Non-Parametric Tests

When the assumptions required by these tests cannot be satisfied, then these are the statistical methods which are designed to take place of the standard parametric tests. They are called non-parametric tests because they do not involve in any of the assumptions which are being made about parameters of the population that is under the investigation. Chi-squared test and Mann-Whitney U test is being included by the non-parametric test.

Normal Distribution

There is a frequency distribution which is symmetrically bell shaped and in which most values lie near the mean of the values at the apex of the curve. Unlike a binomial distribution or poison distribution, it is continuous frequency distribution. In the significance tests one of the assumptions is being made that the values under investigation are normally distributed. As the skewness deviation from the symmetry of the normal distribution is being referred.

North

As the advanced countries sometimes, it is also referred to collectively, the developed world, the first world or MEDC are enjoying the considerable wealth and high standards of living which is being derived from the well-developed secondary and territory sector sectors and also increasingly from the quaternary sector activity. At global scale high rates of literacy and considerable political influence and characteristics also include high levels of energy consumption. Most of the European countries like USA, Canada, Japan, Australia, and New Zealand would be defined as embracing.

Nuclear Power

In the process of nuclear fission, the use of radioactive energy created to generate electric power or to provide propulsion for ships. So, in both the instances to produce the steam the heat is being released by such energy which will also in turn used to drive the turbines. It has been said that about possible radiation hazards and about the safe disposal of radioactive waste. In both instances, the heat released by such energy is harnessed to produce steam which, in turn, is there being public concern, as the use nuclear power is a highly controversial issue. To manufacture nuclear weapons any country that has the capacity is UK, USA, the Russian Federation, China, France, and India.

Nucleated Settlement

It is known as the settlement in which there is a close juxtaposition of dwellings and other buildings, and where there is a well-defined break between the settlement itself and the surrounding countryside. In the history it has been seen that to a variety of circumstances, ranging from association with the open field system, this clustered then may have been a response to the exploitation of highly localized resources from the necessity for defence to an involvement in trade, encouraging settlement to become clustered around nodal points in the developing transport network. Variety of different forms can be shown by the layout of a nucleated settlement which is often determined by the pattern of routeways around, or along, which it grows.

Nuee Ardente

By a violent eruption a catastrophic blast of hot gas, steam and the burning dust is being released and descending the flanks of a volcano as a high-velocity 'incandescent cloud.

Null Hypothesis

Between the two or more samples of a population this is used to determine the significance of the differences being observed. It has been assumed by the null hypothesis that there is no significant difference between the samples and that the observed differences are no more than chance variations. In other words, the null hypothesis is a negative assumption that an effect is not present in the samples. It means all the appropriate test are being applied to all the samples, and as a result the hypothesis may or may not be rejected. If the null hypothesis is rejected, then it can no longer be assumed that there is no significant difference between the samples. If, on the other hand, the null hypothesis is not rejected, then the case is 'not proven.' In short, a null hypothesis cannot be proved to be correct, it can only be shown to be false.

Nunatak

In the Greenland and along the margins of the Antarctic ice sheet a rocky peak projecting above the surface of an icesheet. To the freeze-thaw weathering the nunataks are being subjected, to their precipitous slopes remain largely free of snow and ice owing, and the presence of dark rocks that absorb solar radiation and, by re-radiating heat, promote rapid snow ablation.

Nutrient Cycle

For the plant growth the process whereby, mineral elements are necessary are being taken up from the soil and then returned when the plants shed vegetal matter and or die. By the action of micro-organisms and fungi, releases the elements to the soil he decomposition of the resultant plant litter forming a source of nutrients for future plant growth. In the rainforest of tropical regions, the nutrient cycle is sometimes virtually closed, in which case interruption of the cycle by deforestation and agriculture which will have serious adverse effects on soil fertility. Therefore, to the system usually as the nutrient cycle operates, elements are continually added or it may be lost from it so that, over a period of time, the balance in the supply and consumption of nutrients is maintained.

O

Oasis

By the way of springs or wells a place in a desert where the water is available. It has been aiding that when the water table approaches the surface oases occur.

OAU (Organization of African Unity)

In the year 1963 this alliance was setup with the objectives of furthering African unity through the coordination of a range of policies relating to politics, economics, culture, health, science, defence, etc.; eliminating colonialism from the continent. With the notable exception of the Republic of Africa nearly all African nations belong to the organization.

Obsequent Stream

To the original slope a stream whose course runs opposite slope of the land surface. With the trellised drainage the obsequent streams are particularly associated and in the areas of scrap and vale scenery. By the extension of subsequent the initial pattern of consequent streams is being disrupted and along the outcrops of weak rock, which is leading to the process of capture. From the subsequent the work back headwards frequently along with lines captured consequents. Into the scrap face as it undergoes recession, they may take the advantage of cols.

Occluded Front

In a mid-latitude frontal depression, a type of front is being developed, which is resulting from a more rapidly advancing cold front, overtaking a more slowly moving warm front. It has been said that well above the earth surface the effect is to raise the tropical maritime air of the warm sector. The beginning of the decay phase of a depression is being marked by the formation of an occluded front. It has been noted that there are two types of occlusions:

- warm occlusion – where the advancing cold air is warmer than the cold air ahead of the warm front – in this case, it will ride above the leading cold air;

- cold occlusion – where the advancing air is colder than the cold air ahead of the warm front – in this case it will undercut the leading cold air.

Ocean Currents

For distributing heat between the latitudes there is relatively fast flows of water in the oceans responsible and thereby assisting to maintain the atmospheric HEAT BALANCE. Warm ocean currents flow from the Equator towards the poles and cold oceans from the poles towards the Equator. Ocean currents have a considerable effect on the climates of coastal regions in transferring the temperature, for example, the North Atlantic Drift keeps western Scandinavia warmer than other regions at the same latitude. The movement of water different densities are believed to be important factors and the propagation of ocean currents is still not fully understood but the effect of prevailing winds. In the patterns of ocean currents periodically changes may occur, for example, the ELNINO effect results in the replacement of the cold Peruvian current by much warmer waters.

Oceanography

It is considered as the scientific study of the seas and oceans. With the form of ocean basins, the structure, relief and deposits of their floors, movements of seawater in the form of ocean currents, ocean temperatures and salinity physical oceanography is being considered. The study of life in the oceans is known as biological oceanography.

Occupancy Rate

By dividing the number of people in a dwelling by the number of' habitable' rooms (i.e., normally taken to be bedrooms and living rooms), a measure of housing conditions and overcrowding is being derived.

OD (Ordnance Datum)

This means at the Newlyn, Cornwall sea-level, calculated from hourly tidal observations. In feet (formerly) or metres (on current maps) 'above OD heights on Ordnance Survey maps are expressed.

OECD

In the year 1961 Organization for Economic Cooperation and Development was setup in order to promote social and economic welfare throughout the area of member countries by the international coordination of appropriate policies. With Canada, the USA, Japan, Australia and New Zealand, its membership comprises nearly all the countries of Europe together.

Office Activity

The different part of the tertiary sector is broadly connected with the collection, processing, and exchange of information. Along with this, it has been seen that it is possible to distinguish a number of different categories, such as: company administration; finance (banking, insurance, accountancy, stock-broking, etc.). And the professional services such as legal, real estate, advertising, market research, etc.

There are also some of the public sector activities like local and central government, administration of public utilities, quangos, etc. From the prestigious headquarters of a TNC to the one-roomed premises of a private detective an office may range.

Offshore Assembly

In the third world a form of manufacturing largely encountered involving the assembly of a finished product made from parts and components produced in and imported from MEDCS. There are some of the countries which offer cheap labor by locating the labor-intensive work of assembly, as these businesses are able to maintain a competitive pricing of their finished products, as well as improve their penetration of the domestic markets. In SE Asia, Africa, and Latin America there many Japanese companies who are involved in offshore assembly.

Offshore Bar

On a very gently shelving coastline it has been seen that a bank of sand or shingle developed some distance offshore. Before reaching the beach there is one explanation of its formation is that larger waves will tend to break, thereby eroding the sea bed and throwing up the resultant debris to form the bar. By the longshore drift the bar may be lengthened and stabilized by the formation of sand dunes. Inland from the bar, a shallow lagoon will be formed, which may eventually be filled by mud and SILT, and a SALT MARSH may develop.

Offshore Economy

It has been seen that the outcome of TNCS and other businesses in one country investing overseas, be this by setting up branch plants and also entering into the joint ventures or acquiring equity (stock and shares) in foreign businesses. During the 1980s and 1990s, much of the Japanese economy was moved offshore by these means.

O-Horizon

In the form of litter, fermenting matter and humus the uppermost layer of the soil profile which comprise of organic matter. Into the A-horizon by the earthworms the O-horizon is generally dark brown or black in colour and is slowly incorporated and providing much-needed nutrients and the ability to retain water.

Old-Age Index

By dividing the number of persons over retirement age by the number of people of working age a measure of the 'greying,' 'silvering,' 'wrinkling' or ageing of a population is being calculated.

Old Fold Mountains

It has been seen that prior to the alpine Orogeny of middle tertiary times the fold mountains are being created by the fold mountains. During the Hercynian (late Carboniferous-early Permian times) and Caledonian (late Silurian-early Devonian times) orogeny's the old fold mountains of the British Isles, Europe and elsewhere were created. It has been noted that the structural features of old fold mountain are quite similar to the young fold mountains.

Though, to immense amounts of erosion the former has been subjected, in the course of which they have been reduced to peneplains and owing to their great age. The episodes of uplift and rejuvenation are also being experienced by them and now often of plateau-like form, with broad interfluves separated by deeply incised river valleys. In some instances, old fold mountains were inundated by marine transgressions and became covered by younger sedimentary rocks. Following uplift, these have been stripped away (except around the margins), revealing exhumed surfaces of erosion. The drainage systems developed on the younger rocks have been superimposed on to the older structures to give strikingly discordant patterns.

However, the landforms of old and young fold mountains are not wholly contrasting. In many cases (e.g., the European Alps and the Scandinavian highlands), similar processes have operated; thus, Pleistocene glaciation has produced cirques, aretes and glacial troughs in both regions.

Oligopoly

For a large proportion of output and employment an economic activity in which a small number of firms are accounted, e.g., the motor vehicle industry.

Oligotrophic Lake

By a relative lack of dissolved minerals and a generally low nutrient status a lake that is being characterized. In the upland granite regions, there are many deep lakes, such as parts of Scotland and Norway, are oligotrophic. To the considerable depths the water lacking in microscopic algae, is clear, and sunlight penetrates which allowing the growth of aquatic plants. Therefore, it has been seen that because of the lack of nutrients the flora and insect life are impoverished (with certain exceptions, such as mayflies), and salmon and trout are the only large fish species present.

Figure 58. Oligotrophic water body.

Source: Image by flickr.com.

Onion-Skin Weathering

From a large boulder or the exposed face, the detachment of the curved sheets of a rock is the result of unloading, surface expansion and contraction related to temperature changes, or pressure from within the rock exerted by hydration of minerals or salt crystal growth.

OPEC (Organization of Petroleum Exporting Countries)

It has been formed in the year 1960 with the aim of unifying the oil policies of member countries and of determining the best means of safeguarding their interests. There some countries which have their membership like Algeria, Ecuador, Gabon, Indonesia, Iran, Iraq, Kuwait, Libya, Nigeria, Qatar, Saudi Arabia, the United Arab Emirates, and Venezuela.

Opencast Mining

It is considered as the form of extensive excavation, only practicable where the mineral deposits lie at or near the surface. By the large-scale machinery the overlying material (overburden) is removed and seems or deposits thus revealed are then quarried. It has been said that the American terms are strip mining or open cut. For working coal, brown coal, and iron ore it is a technique which is being used particularly.

Optimal City Size

A city said to be ideal only when it can maximize the benefits, minimize the cost and generate the greatest net benefit. There are certain trends in both costs and benefits which is accompanying the increasing city size and the notion is based on this only. It has been noted that the cost curve is being depicted as the U shaped and the benefit curve as S-shaped. Therefore, on a graph, there are certain key points that may be identifies along with the city size axis.

PI is the threshold at which benefits exceed costs (i.e., the onset of net benefit); P2 is the city size that incurs least costs; P3 is the city size that attains the greatest net benefit (i.e., largest margin of difference between costs and benefits); P4 is the city of greatest benefit; while P5 is the point at which costs once again begin to outweigh benefits. So, PI and P5 might be specified as the desired lower and upper limits to city size, while P2, P3 and P4 might be claimed as indicating optimal city sizes.

Optimal Location

In the economic geography this term is much used during the normative phase and also variously interpreted to mean the best location from an economic viewpoint, be it the least cost location, the location yielding maximum benefit or the greatest profit, or the location that best satisfies the particular needs of a given activity.

Optimizer Concept

For the maximum productivity and profit in the conduct of business this is the concept which holds that the entrepreneur or decision-maker always strives. During the mid-20th century and is an integral part of neoclassical economics it dominated the economic geography.

Optimum Population

To the differing interpretations a concept that is open is known as optimum population. In any of the area in relation to the possibilities and resources of the environment and enjoy a reasonable standard of living it might be defined as the number of people who can live most effectively. There are many factors by which the number of people will be influenced including the nature and productivity of the mode of production, the level of technology, etc. In a given area size of population allows the maximum utilization of resources, and which achieves the greatest output per head and the highest living standards.

Order

This relates to the hierarchic status of individual goods, services, and central places in the central place theory. It has been seen that order is largely determined by their threshold and range value in the case of goods and services, while with central places it is allied to position in the central place hierarchy.

Ordinal Data

In terms of their relative importance the data is being presented rather than their absolute value. For instance, into fine, medium, and coarse categories the classification of bedload material has been done or the grouping of settlements into the sizes it would constitute.

Organic Farming

It is considered as the farming in which there is no use chemical fertilizers, pesticides, herbicides, and fungicides. Apart from this from fish and bone meal the use is made of animal and green manures, and fertilizers. A smoothing boom is being experienced by the organic farming, along with the growing concern about chemicals entering the human food chain and adversely affecting health. It is also thought to be much less harmful to wildlife and the environment. Because yields are generally lower than with modern intensive farming, products are more expensive. But it seems that more and more people are willing to pay that little bit more for what is seen as 'safe' food.

Orogeny

It has been noted that a major episode of mountain building, involving powerful compressive folding and faulting of previously formed sediments which is being followed by the uplift and erosion. Since the Pre-Cambrian have been the Caledonian, the main orogeny's is affected.

Outlier

By the older rock an outcrop of the younger rock completely surrounded. As a result of differential erosion of escarpment faces the outliners are commonly formed. Into the scrap parts systems of stream valleys are eroded back and parts of the caprock of the latter may become isolated.

Out-of-Town Shopping Center

At the edge of the built-up area or just beyond it a large retailing complex is located. There are some factors which encourage the such type of development include relatively low cost of land, room for the expansion and accessibility from a much wider area than just the nearby town or city. In the UK there are some of the largest centers which do not contain just shops but a whole range of leisure amenities in order to encourage a day out for the family.

Outport

To provide service to an established port that is proving in some way inadequate a subsidiary port is being developed, often at the month of an estuary. For the modern shipping outports have been developed, as shipping has grown in size and the original port has become increasingly difficult to access. As estuaries have become progressively silted up as passenger traffic has required speedy transport by rail or road to its destination, instead of proceeding slowly by ship further up the estuary, where large areas of land have been required for the development of new port facilities.

Output

By a firm an industry, a region or country the quantity of goods produced or services provided during a given period

Outwash Plain

By the meltwater streams there is a gently sloping area which comprise of sands and gravels deposited and flowing from the stationary margins of a glacier or ice sheet. There are large outwash plains which are related to the contemporary glacial conditions and are found in Iceland and are called sandur. By the streams these are being built up that migrate slowly across the plain from one side to another, indeed only by a major glacial outburst the outwash plain as a whole will be inundated.

Overcapacity

For its products there is excess of productive capacity that is likely to arise in an industry subject to constant fluctuations in the demand, at the time when the level of demand falls below the level of output, i.e., surplus production.

Overflow Channel (Also Spillway)

From the glacial lake a channel is being cut out by the water draining. There are some of the features that are being interrupted as overflow channels are now believed to have resulted from subglacial stream erosion.

Overgrazing

On land putting too many animals to graze in farming, so that vegetation cover is gradually destroyed. This removal of the protective mat of vegetation can lead to serious soil erosion in an area of uncertain or short seasonal rains.

Overland Flow

By the vegetation the proportion of total rainfall that is not intercepted and does not infiltrate the soil, but runs over the ground surface as sheet or rill wash. With the rain fall intensity the volume of overland flow increases and usually at a maximum during convectional storms in semi-arid and tropical regions. At the time when rainfall exceeds the infiltration capacity of the soil or, more commonly, when the pore spaces of the soil are saturated and unable to absorb additional water overland flow occurs. Overland flow represents very rapid transfer of water, and is associated with the rising limb and peak of a storm hydrograph. It is most likely to occur in areas of steep slopes, thin soils, impermeable bedrock, lack of vegetation, frozen soil, intense rainstorms, etc.

Overpopulation

There is a condition in which the population of an area exceeds the required number of population and therefore by the available resources, technology, and mode of production it cannot be supported adequately. So, for this a consequence is that the standard of living is low and decline still further. With the dense population overpopulation is not necessarily equated, for instance, cities with high density are not over populated while some mountains can be over populated and desert areas as well.

Overspill

From the densely crowded urban area the surplus population and employment that can no longer be effectively accommodated and maintained there, and which removes either voluntarily or in response to statutory plans. In the new suburbs such decentralizing people and firms relocate and in nearby less congested TOWNS, or remove to planned overspill reception schemes located further afield.

Overurbanization

In the third world countries a state of affairs thought to exist, where it is claimed that the degree of urbanization is too high relative to the level of economic development. From such a situation the particular problem is being resulted and it has been said that substantial rural-to-urban migration creates a large urban labor force that cannot be fully employed by the secondary and tertiary sectors at their current levels of development. Therefore, it will result in unacceptable levels of unemployment, from which spring such things as poverty.

Oxbow Lake

By the breaching of the meander neck a crescent-shaped lake or marshy area, occupying a former river meander that has become cut off. It has been seen that oxbows are the characteristics of the largest flood plains, as in the Severn and Trent valleys in Britain, and the Mississippi valley in the USA.

Oxidation

It is considered as the common process of chemical weathering where the rock minerals get combine with oxygen, usually dissolved in infiltrating soil water. Reduction is the reverse process of oxidation.

Ozone Layer

In colour blue ozone which is the form of oxygen found in greatest quantities at a height of 20–25 km in the Earth's atmosphere. Through the absorption of ultraviolet radiation by oxygen ozone is believed to result from photochemical changes. It has been seen that ozone layer act as a filter which protect the lower atmosphere and ground surface from potentially harmful ultraviolet rays. In the ozone layer the concentrations of the ozone are being seriously depleted as per the recent research, and largely as a result of the build-up in the atmosphere of chlorofluorocarbons (CFCS) from sources such as domestic refrigerators and aerosol spray cans. In the layer so-called holes have been discovered, first over Antarctica and, more recently, over the North Pole.

In the urban areas ozone is harmful pollutant in summer at the lower levels and weather forecasts now frequently include references to 'ozone pollution'.

P

Pacific Rim

There are some of the countries that bordering the Pacific Ocean like Japan, China, Asian Tigers, and other countries of SE Asia, together with the USA, Canada, Mexico, and the countries of Central America. There is Australia, New Zealand and the American countries of Colombia, Ecuador, Peru, and Chile in the Southern Hemisphere. During the last quarter of the 20th century that part of the Pacific Rim lying in the Northern Hemisphere now challenges as the leading global, sunrise region because of the economic growth.

Pacific Ring of Fire

Other features of tectonic and seismic activity the concentration of volcanoes around the margins of the Pacific Ocean corresponding with plate margins.

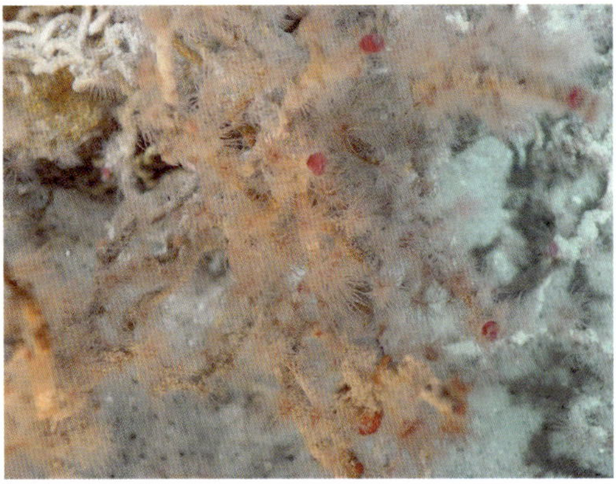

Figure 59. Pacific ring of fire expedition.

Source: Image by flickr.com.

Pacific-Type Coast

To the coastline as a whole, it is a type of coastline that is formed where the trend of mountain ridges and valleys is parallel e.g., the coast of British Columbia. By the lines of islands and straits a coast tends to be straight and regular is being characterizes.

Package Tourism

It has been said that in which holiday travel and accommodation are put together as a relatively cheap package is the sector of tourism industry. From the economies of scale, it is aimed at the mass market and the most popular tourist destinations.

Paleoecology

During the Pleistocene and post-glacial periods, the science of the reconstruction of past ecological conditions from floral and faunal remains preserved in deposits such as PEAT. About former climates and climate change the reconstruction of former vegetation conditions can provide the basis for inferences.

Paleolithic Period

Coinciding with the greater part of the plies-TOCENE Ice Age, the old stone age is considered as the earliest period of human prehistory. To the various glacial and interglacial phases, a series of culture periods has been distinguished and related. It has been said that earliest implements were roughly fashioned were considered as flints, but gradually these became more highly fabricated, and at some stages bone was used. Specially in France cave paintings is the significant feature.

Paleomagnetism

In the rocks there is the phenomenon of the fossil magnetism. By the two related forces the needle of a normal magnetic compass is influenced. It has been said that it will point in the direction of the magnetic North Pole and if there will be a situation that is provided freely then will dip at an angle to the horizontal *(inclination),* the precise amount of dip depending on latitude. Therefore, to become vertical at the pole at the Equator, dip is zero, and will increase progressively.

Thus, at the time when igneous rocks cool, magnetic minerals (comprising iron) will, as they crystallize out, behave like freely suspended magnetic needles, thus recording the direction and distance from the magnetic pole at the time of cooling; the magnetism forms a 'fossil' record for that time. Similarly, at the time of deposition magnetic grains in the sedimentary rocks becomes oriented lithified to form sandstone or shale, and though in a weaker form the magnetism will again be preserved.

Over the geological time studies of the residual magnetism of rocks provide information as to changes in the Earth's magnetic field. Many times, this field has been reserved, as if the North and South Poles exchanged places. Paleomagnetism has been used to provide evidence supporting the concept of continental drift and current theory of plate tectonics.

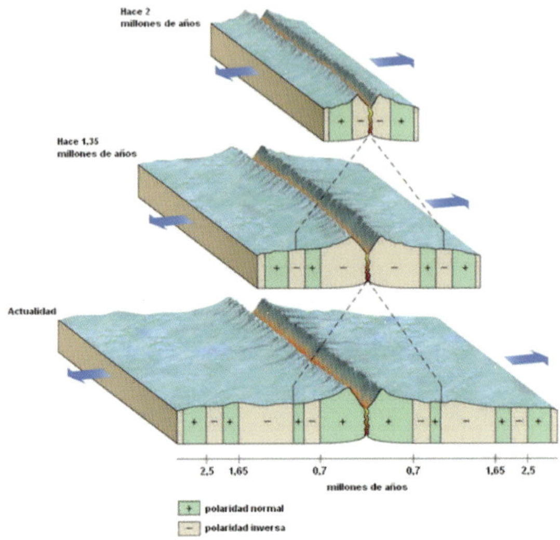

Figure 60. Paleomagnestisme.

Source: Image by Wikimedia.

Palaeotechnic Era

During the technological developments the first phase of the industrial revolution encouraged the growth of heavy industries, transport networks were still relatively primitive; markets were highly localized, and the urban settlements were comparatively compact. From c 1752 to c 18870 the palaeotechnic era lasted.

Pallid Zone

Between the overlying mottled zone and the underlying partially decomposed bedrock a widely identified weathering horizon lying between in the tropical regoliths. At a thickness of 60 m or more the pallid zone may reach but it is usually less than 25 m. As the pallid zone the horizon is being referred because of its very pale appearance when exposed in section.

Palsa

In the peat bogs and periglacial regions, a conical or elongated mound, containing a lens of ice has been occurred. Palsa are several metres in height, and up to 100 m in diameter. Silt is being comprised by the palsa which is underlying the PEAT and which freezes as the surface cold penetrates the bog; as a result, the peat layer is uplifted and domed.

Pampa

In the Argentina and parts of Uruguay, America there is an extensive grassy plain. With the temperate grassland the term is sometimes regarded as synonymous.

Pandemic

A disease, such as AIDS, affecting large numbers people over much of the world.

Parabolic Dune

In the plain there is a type of dune, and the characteristic of sandy coastlines which is curved and with 'tails' that are pointing upwind. In terms of stabilization at each end by marram grass parabolic dunes have been explained, while the central higher part of the dune has continued to migrate inland under the influence of aeolian transport of sand from the windward to the lee slope.

Figure 61. Parabolic dune.

Source: Image by flickr.com.

Parallel Drainage

The drainage pattern in which the rivers flow almost parallel to each other is known as parallel drainage. The small and swift rivers originating in the Western Ghats and discharging their water into the Arabian Sea provide a good example of parallel drainage pattern in India.

Parallel Sequence

A parallel sequence is a sequence that runs parallel to operations in the standard sequence. The start of the parallel sequence corresponds to the start of the branch operation in the standard sequence. The end of the parallel sequence corresponds to the end of the return operation.

Parameter

An arbitrary constant whose value characterizes a member of a system (such as a family of curves). Also, a quantity (such as a mean or variance) that describes a statistical population.

Parametric Tests

Parametric tests are designed for idealized data. In contrast, nonparametric tests are designed for real data: skewed, lumpy, having a few warts, outliers, and gaps scattered about. Nonparametric methods are workhorses of modern science, which should be part of every scientist's competence. Beyond that, they are very valuable for learning data literacy because they encourage the student to gain a tangible "feel" for the data they are examining. Carrying out nonparametric tests may involve reordering the datapoints into ranks, pairing them up across groups, flipping coins to determine outcomes, or shuffling subjects or samples among different groups.

Participation Rate

The participation rate refers to the total number of people or individuals who are currently employed or in search of a job. The participation rate is also important in understanding the unemployment rate in the economy.

Particle-Size Analysis

Particle size analysis, particle size measurement, or simply particle sizing is the collective name of the technical procedures, or laboratory technique which determines the size range, and/or the average, or mean size of the particles in a powder or liquid sample Particle size analysis is part of particle science and its determination is carried out generally particle in technology laboratories.

The particle size measurement is typically achieved by means of devices called Particle Size Analyzers (PSA) which are based on different technologies, such as high definition, analysis of Brownian motion gravitational settling of the particle and light scattering (Rayleigh and mien scattering) of the particles. The particle size can have considerable importance in a number of industries including the chemical, food, mining, forestry, agriculture, nutrition, pharmaceutical, energy, and aggregate industries.

Passive Glacier

A glacier with a very low rate of flow, owing to a combination of gentle gradient (e.g., On a plateau surface) and small inputs of snow and ice, and limited melting.

Pastoralism

A type of farming concerned mainly with the rearing of livestock, whether for meat, milk, wool or hides. It may be extensive (e.g., The sheep farms of Australia, the ranches of USA) or intensive (e.g., The dairy herds of Europe). It may be nomadic and of low technology (as practised in parts of Africa and the Middle East) or highly scientific (as in Europe) with careful breeding programs, close monitoring of feeding and livestock performance, as well as involving the use of sophisticated machinery.

Pasture

An area of land covered with grass used for the grazing of domesticated animals, as distinct from that which is mown for hay (meadow), although the same field may be used for both purposes at different times of the year. Some pasture may be 'natural' (e.g., An ALP or mountain pasture), but most areas of grazing are usually improved in some way (by liming, fertilizing, draining and periodic reseeding).

Paternoster Lakes

A series of lakes in a formerly glaciated valley, separated from each other by morainic deposits or rock bars, but linked together by streams. The term is derived from the resemblance of such lakes, when viewed in plan, to a string of beads (paternoster = a bead in a rosary).

Patterned Ground

A surface and sub-surface pattern of course and fine debris, most evident in areas with little or patchy vegetation cover, which is produced by lateral sorting processes, most frequently in a periglacial environment. Patterned ground is highly variable in character, and includes such features as stone polygons, fissure and ice wedge polygons, and stone stripes.

Peak Land-Value Intersection (Point)

The highest point on the urban land-value surface; i.e., the point within the town or city where land values are greatest. It usually occurs at a major traffic intersection within the CBD, and its location is marked by the most prestigious shops.

Peat

An organic deposit comprising partially decomposed plant materials that have accumulated in a waterlogged (anaerobic) environment under conditions of impeded drainage (e.g., on flat areas or valley floors) and/or during prolonged wet periods over the landscape generally. Peat is found in cold or temperate climates. In the latter it is often a relict formation, developed under wetter conditions in the recent past. Peat is widely used for fuel (domestically, or in peat-fired power stations) and as a conditioner for garden soils. This has led to the widespread destruction of peat bogs (a distinctive and irreplaceable habitat for certain types of flora and fauna).

Q

Quadrat Analysis

A statistical technique used in the analysis of point patterns. The area of a point pattern is divided into equal-sized cells (usually grid squares) and the frequency of occurrence of points in each cell is determined. The observed frequency distribution is then compared with some theoretical or 'expected' pattern, thereby establishing the precise character of the point pattern under investigation (i.e., Whether it is uniform, random or clustered) and of the processes producing it.

Quadrat Sampling

A method of sampling usually employed to reduce investigation of a large area to more manageable proportions. The study area is subdivided into a scheme of smaller units of equal size, this subdivision being most easily achieved by superimposing a grid of an appropriate scale. The investigation can then be directed at a sample of those grid-square areas. That sample may be selected either on a random or on a systematic basis.

Quality of Life

A complex concept concerning the general state or condition of a population in a given area. Undoubtedly, it has an important psychological dimension that takes into account such states of mind as satisfaction, happiness, fulfilment and security (sometimes referred to as social satisfaction). It also has a physical dimension that embraces such criteria as diet, housing, access to services, and safety. Other aspects include considerations such as social opportunity, employment prospects, affluence and leisure time. Some hold that it is synonymous with well-being and others, however, interpret it as a particular expression of well-being, distinguished by an emphasis on the amount and distribution of public goods and services. It is a term commonly and loosely used in the literature of welfare geography.

Quangos

An acronym derived in the 1970s from the abbreviation of the name given to bodies variously referred to as: quasi nongovernmental organizations; quasi autonomous non-government organizations; quasi autonomous national government organizations. In effect, these three titles refer to the same thing, namely a semi-public administrative body, outside the civil service, but financed by and having members appointed by government. In Britain, examples of quangos would include the BBC, the British Tourist Authority and the Police Complaints Board.

Quantitative Techniques

Statistical methods for abbreviating, classifying and analysing numerical statements of fact in any investigation and making clear the relationships between phenomena. Quantitative techniques fall into two broad classes: descriptive statistics and inferential statistics.

Quarrying

In physical geography, a generic term sometimes applied to the erosion of a sea- or river-cliff by the combined forces of hydraulic action and cavitation. In glaciation, the term has been used to describe the effect of extreme water pressures recorded within the upper part of permeable bedrock beneath a glacier, believed to be capable of causing the rock to break apart to produce a jagged and irregular surface.

Quartiles

The percentiles that divide a distribution into four equal parts (i.e., The 25th, 50th, or MEDIAN and 75th percentiles). The 25th and 75th percentiles are known respectively as the lower quartile and the upper quartile.

Quartzite

Very hard rock, consisting of quartz grains bonded to each other or cemented by secondary silica. Quartzite is particularly resistant to chemical weathering.

Quaternary Sector

A relatively recently recognized economic sector comprising all those personal services that require high levels of skill, expertise and specialization, e.g., education, r&d, administration and financial management. Unlike the tertiary sector, this sector is concerned with people and information rather than goods.

Questionnaire Survey

A data collection method involving questions and answers, and used for obtaining information about the way people live and what their views and opinions are on particular issues. Such surveys may be conducted in the street, door-to-door or by post and telephone.

Quick Clay

A clay-rich deposit, comprising fine glacially abraded particles laid down in water, which when subjected to stress change from a solid to a liquefied state, and then undergoes rapid flow.

Quick Flow

The peak on a hydrograph showing run-off related to an individual rainstorm.

Quota

A limit imposed on the quantity of goods produced or purchased. Import quotas can be used to restrict the purchase of goods from foreign origins, while export quotas have been used to stabilize the export earnings of countries producing primary products by restricting supply and thereby sustaining prices. Quotas may also refer to the minimum level of production required in planned economies.

R

R&D (Research and Development)

A vital input to a whole range of industries, particularly high-technology industries. The research is undertaken with a specific commercial aim in mind, while the development refers to the work of exploiting that research in either new processes or products, or the improvement of existing processes and products. Expenditure on R&D is crucial if the firm is to remain competitive, particularly in the sense of being technically innovative and of playing a leading role in the commercial application of new technology. R&D is regarded by many as the root cause of technical progress, innovation and economic growth.

Radburn Layout

A style of residential layout pioneered at Radburn, New Jersey (USA) between 1928 and 1933, subsequently adopted and adapted in the planning of post-war housing areas in Britain, particularly in newtowns and expanded towns. Inspiration for the layout came from the ward idea incorporated in Howard's garden city plans. Its main features include the segregation of pedestrian and vehicular traffic, 'turned around' housing fronting on to amenity space and gardens, and with vehicular access to the rear, loop roads and Culs-de-Sac. In the British post-war new towns, the Radburn principles were clearly evident in the detailed plans of neighborhood units.

Radiation Fog

Fog that is formed at night and early in the morning under conditions favoring free radiation of heat from the earth's surface. The latter becomes cooled, and the temperature of the overlying air is reduced by conduction of heat to the cold ground surface. If the air is moist, dew-point will be quickly attained, and numerous tiny water droplets, with a very slow settling rate, will be formed by condensation within the cooled air layer. Radiation fog is associated particularly with anticyclonic conditions (clear skies permitting maximum nocturnal radiation) and light breezes (needed to spread the cooling upwards from the ground).

The presence of 'water-attracting' nuclei such as particles of dust, salt and other chemicals may allow condensation to occur before relative humidity reaches 100%. Radiation fogs are often thickest over moist valley bottoms, where the process of formation is intensified by drainage of cold air from adjacent slopes. In summer the fog will be dispersed quite rapidly as the sun raises air temperatures above dew-point, but in winter it may persist all day, or even intensify over a period of days.

Radio-Carbon Dating

A technique development containing carbon. Carbon-14, a radioactive isotope of carbon with a half-life of 5570 years, is formed in the upper atmosphere, subsequently oxidizes to carbon dioxide, and becomes absorbed by living plants. Following the plants' death and the preservation of their remains within sediments under favorable circumstances (buried organic soils, PEAT formations, etc.), the Carbon-14 content diminishes over time at a known rate. From this, the age of the organic remains (and thus the deposit in which they lie) can be calculated within fairly close limits.

Raindrop (Rain Splash) Erosion

The displacement of soil particles by the impact of large raindrops, particularly under conditions of bare soil affected by intense convectional rainfall, for example, in semi-arid environments. The raindrops have the effect of detaching individual particles, which are then moved downslope. As they do so, the finest particles are forced into voids, reducing infiltration, and increasing surface run-off. Under forest conditions, the direct impact of raindrops will be reduced by interception, but water dripping from branches and leaves will still have some effect. In Savanna regions, the onset of heavy rains at the end of the dry season will lead to raindrop erosion on baked, unvegetated surfaces.

Rainfall Intensity

The amount of rain that falls in a given time period, usually expressed in millimetres per hour. Rainfall intensity is thus a measure of how 'heavy' the rain is. Maximum values, which may be as high as 100–150 mm per hour, are usually associated with: tropical environments, where prevailing high temperatures and rapid evapotranspiration result in large values of absolute humidity; intense convective uplift leading to very rapid cooling and condensation of moist air over restricted areas. Rainfall intensity has important implications for landform processes such as rain wash erosion on slopes.

Rural Turnaround

The revival of more accessible rural areas, previously suffering from depopulation, brought about by the process of counter urbanization. The 'flight from the city' started some time ago with the dispersal of retired people anxious to spend their last years in calmer, cheaper and more attractive locations. They are now followed by growing numbers of commuters, and in turn by various activities such as R&D, offices and health clinics.

Rurban

A composite adjective introduced into geographical literature, and applied to the indeterminate and transitional condition between country and town, between the rural and urban states.

Ruware

A low, rounded, and often elongated exposure of unweathered rock (sometimes referred to as a rock pavement or whaleback) rising a few metres above the surrounding plain. Ruwares are common in Savanna regions.

S

Sahel

An E-W zone along the southern margins of the Sahara Desert in Africa. Although the mean annual rainfall should be as high as 500 mm in places, the Sahel is subject to severe droughts (like those of the early 1970s and 1980s), bringing crop failures, death of livestock and famine to the inhabitants. There are reasons to believe that these droughts are partly a natural phenomenon, related to long-term climatic changes, but their effects have been emphasized by human activities such as overstocking, destruction of natural vegetation, etc.

Saline Soil

A type of SOIL characterized by a high salt content. The salt (mainly sodium chloride and sulphate) is derived either from the evaporation of groundwater lying just beneath the soil surface, or is drawn towards the surface in solution by the action of capillarity. In arid regions where rates of evaporation are high, constant irrigation can lead to considerable increases in soil salinity.

Salinization

The processes whereby sodium, potassium and magnesium salts become concentrated in the SOIL. Natural salinization is characteristic of many desert regions, where evapotranspiration exceeds precipitation, leading to the capillary rise of salt solutions from groundwater, and precipitation of salt crystals either within the upper soil layers or as a surface crust. Salinization can also result from excessive IRRIGATION in hot, dry climates; the resulting salt-pan is poisonous to agricultural crops.

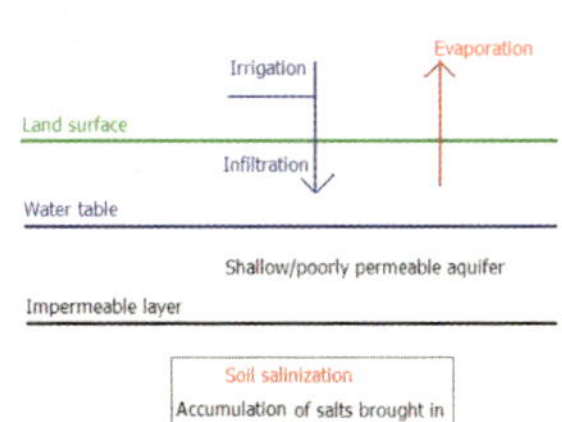

Figure 62. Salinization process.

Source: Image by Wikimedia.

Salt Marsh

A coastal marsh, formed within protected estuaries or in the lee of coastal spits, bars and sand dunes. The salt marsh is gradually built up by the deposition of fine silt and mud, which settles out at periods of slack tide. These sediments are mainly derived from the suspended load of rivers entering the sea, but some may come from the erosion of clay cliffs. When the level of the marsh has been raised sufficiently, it is colonized by plants that can withstand periodic inundation by salt water.

As the vegetation cover extends, the rate of sedimentation accelerates, and the marsh eventually reaches a level at which it is flooded only by high spring tides. Noteworthy features of fully developed salt marshes are pans (small unvegetated areas) and creek systems (channels, by way of which salt water enters and drains away from the marsh). People frequently reclaim mature salt marshes by the construction of embankments that exclude the sea, thus providing grazing for animal.

Salt Weathering

A type of weathering, now recognized as particularly important in hot deserts but also occurring in tropical humid climates, involving a chemical process but having a physical effect on rocks. Salt weathering results from the crystallization of supersaturated solutions of salts occupying pore spaces and cracks.

Figure 63. Salt weathering.

Source: Image by Wikimedia.

Saltation

A process of transportation in river, involving the continual 'jumping' of small particles (usually sand grains) along the channel floor. Any particle lying on the river bed will impede water flow, and hydraulic pressure will be built up on the upstream side of the particle. In time, the stress will exceed the resistance of the particle (determined by its weight and shape) and it will then be set in motion, and may be thrust up into the mass of moving water. However, unless the particle is very small and light, gravity will cause it to fall back to the bed, where it may help to dislodge another particle. The term is also used to describe a similar type of transportation by winds.

Sampling

The essence of sampling lies in the fact that a large number of items, values, individuals or locations (i.e., a statistical population) may, within specified limits of statistical probability, be represented by a small group of items, values, etc. (i.e., By a sample) selected from that original population. A resort to sampling is necessary in many research topics, principally because of: the number of variables involved; the vast dimensions of the phenomena under investigation; the constraints of time, effort and labor. The key to success in sampling lies in adopting a procedure that permits the drawing of satisfactory and valid conclusions about the parent population from a sample of minimum size. There is a variety of sampling methods.

Sampling Error

The difference between the sample statistic of a population attribute and the actual parameter. It occurs because sampling is almost inevitably biased either to over- or underestimate the attribute being investigated. In random sampling limits can be calculated for the sampling error.

Sand

Small grains, usually composed of quartz, with a diameter of between 0.06- and 2-mm. Sand is divided into fine (0.06–0.2 mm), medium (0.2–0.6 mm), coarsc (0.6–1 mm) and very coarse (1–2 mm). Sandy soils comprise a mineral content of sand grains in excess of 90%.

Sand Dunes

Generally, smoothly curving features resulting from the action of the wind on deposits of SAND, most commonly found in desert environments or at the coast. Sand dunes may take many forms, including crescents and linear ridges. They are very dynamic features whose shape and dimensions reflect the varying directions and strengths of the wind. Over time, vegetation (such as marram grass at the coast) may be able to colonize the dunes and a vegetation succession takes place. Coastal sand dunes result from sediment deposited by the sea being reworked by the wind. Extensive sand dune complexes can indicate a relative fall in sea-level.

Figure 64. Sand dunes.

Source: Image by libreshot.com.

Sandstone

A sedimentary rock comprising mainly grains of quartz (though feldspar may also be present) that have been compacted by pressure and cemented by calcareous, siliceous or ferruginous minerals. Sandstone is thus a more coherent and resistant rock than sands. It is generally regarded as being a porous and permeable rock and, in some parts of Britain, forms an important aquifer.

Sandur

An Icelandic term for a large glacial outwash plain. In Iceland, these features are very extensive indeed and are criss-crossed by braided rivers. Periodically, they are swept by glacial outbursts.

Seedbed Location

An area where conditions favor the birth of new industrial enterprises. Typically, seedbed locations are to be found in declining inner-city areas where there is an abundant supply of cheap premises that can easily be converted into small factories.

Segmented Economy

An economy comprising a variety of firms, ranging in size from TNCS to small backyard workshops.

Segregation

In an ecological sense, the spatial separation of different groups of people and different functions into distinct areas, e.g., The emergence of distinct social areas and different land use regions within the built-up-area of a town or city. Segregation results partly from the repelling force that operates between different activities (and social groups) and partly from the mutual attraction that exists between similar activities

Seismology

The scientific study and interpretation of earthquakes. Hence seismologist (a student of earthquakes), seismic (a descriptive term for earthquake processes) and seismograph (an instrument for recording earthquake shock-waves).

Selective Migration

The migration of a particular group or type of person, as by young people seeking work or by a persecuted ethnic minority. At an international level, there are governments that pursue policies of selective migration. For example, entry into Australia is carefully controlled, the issue of residential visas being largely restricted to people who command specified job skills and who fall within a prescribed age range. Such policies clearly represent a form of discrimination.

Shale

A sedimentary rock formed by the compaction of fine muds and characterized by laminations (thin layers that easily split apart). Shale is generally a week, impermeable rock that is easily eroded and tends to form areas of low relief.

Shanty Town

An area of substandard housing, often occupied by squatters and found mainly in third world cities. Usually constructed either at the city margins or on difficult ground (e.g., steep slopes, areas prone to flooding) within the city, hitherto avoided by the built-up area. While shanty towns are sometimes referred to as squatter settlements or shanties, they frequently have local names, such as the barricades of Peru, the favelas of Brazil and the villas miseries of Argentina.

T

Taiga

A Russian term for the coniferous forest belt extending across the northern part of Eurasia.

Talus

An accumulation of angular fragments on a slope. Talus may comprise a relatively thin veneer of debris, resulting largely from in situ weathering of the underlying rock, or a thick deposit of fragments that has collected at the base of a free face.

Tariff

A duty or tax charged by a country on its imports from other countries; a customs duty. The duty may be imposed as a percentage of the value of the goods or as a specific amount per unit of weight or volume.

Tariff Barrier

The use by a country of a tariff in order to protect its own industries from the competition of foreign producers. By imposing a barrier in the form of a high tariff, the price of an imported commodity is increased and thus that commodity becomes less competitive in the market-place. Tariff barriers may also be used by a country to reduce the total volume of imports, especially where there is a balance of trade deficit.

Taxonomy

The scientific classification of features according to general principles and laws, e.g., the Linnaean classification of plants, Kappen's classification of world climates, a classification of towns on the basis of function or central place status.

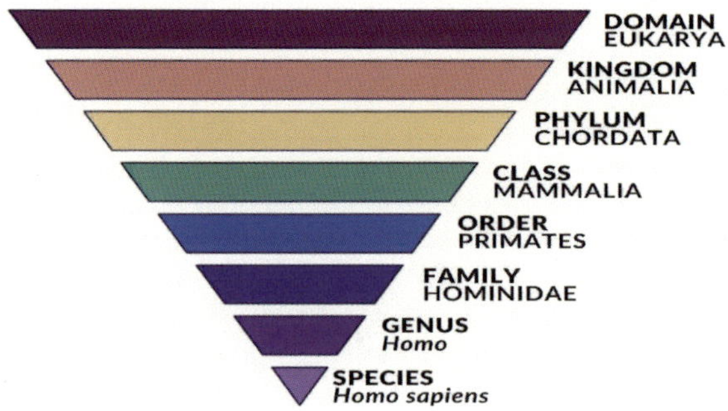

Figure 65. Taxonomy of living things.

Source: Image by examples.com.

Technology Transfer

The passing on of innovation and technological advances from one country to another. One of the potentially most important of such transfers is from MEDCS to LEDCS in the context of AID. However, transfers also take place between MEDC, as for example, the transfer of Japanese work practices to European branch plants. In both geographical settings, TNCS are important promoters of such transfers.

Technology Treadmill

This refers to the vicious circle in which many small farmers particularly in the third world find themselves caught. In order to keep up with larger producers, they are forced to buy either more land or the technology needed to raise productivity. Gradually they are drawn into debt and the only way to relieve that debt burden is to raise productivity still further by buying in still more land or technology. The farmer is therefore stuck on a treadmill from which it is extremely difficult to escape.

Tectonic

A term describing movements within the Earth's crust, resulting in uplift and depression, lateral sliding, warping, folding and faulting. Landforms produced directly by such movements are sometimes referred to as tectonic relief.

Tenant Capital

The equipment, such as seed, livestock, machinery and fertilizers, supplied by a tenant in an agricultural system.

Terminal Costs

The costs of loading and unloading freight at the points of origin and destination, and at break-of-bulk points; a component of transport costs.

Terminal Moraine

Also referred to as end moraine, a ridge of boulders, gravel, sand and silt formed at the terminus of a glacier or ice sheet. In many instances the constituent debris consists not only of subglacial particles but also of supraglacial and englacial sediment, exposed by surface melting of the ice. Terminal moraines are usually asymmetrical in cross-section, with a steep ice-contact face on the side next to the glacier (proximal slope) and a gentler slope away from the ice (distal slope). In the case of valley glaciers, terminal moraines may merge with lateral moraines.

Terminal Velocity

The rate of fall attained by a particle passing through a liquid or gas. The precise velocity is determined by the size and weight of the particle (influenced by the gravitational pull), and the resistance to movement afforded by the liquid or gas. In the atmosphere the terminal velocity of raindrops and hailstones may be countered by rapidly rising updraughts of air; this is an important factor in their subsequent enlargement.

Terra Rosa

A reddish soil developed on limestone in areas of Mediterranean climate; the colour of the soil is due to the presence of iron hydroxides. Terra rosa largely comprise insoluble particles within the limestone that have been exposed as the surface has been lowered by carbonation. It is thus essentially a residual soil.

Terrace

A bench-like feature in the landscape, delimited at the 'front' by a relatively steep drop to lower ground and at the 'back' by a pronounced bluff. Terraces are either erosional or depositional in origin, and can be formed by a variety of processes (i.e., wave erosion may produce a platform that, if subsequently upraised, becomes a marine terrace).

Terracette

A very small terrace, up to 30 cm in width and often unvegetated, on a steep grassy slope (usually at an angle of 30° or more). Terracettes occur in parallel series, running approximately along the contours of the hillslope, with a spacing of about a metre or so. It has been suggested that they result from the rupture of the turf mat by active soil creep. However, there is little doubt that, in many areas, no matter what the process of initiation, terracettes have been greatly exaggerated by trampling animals.

Territorial Justice

A concept encountered in welfare geography that involves relating the level of government spending on such matters as housing and social services in a given area to the needs of that area. Research has demonstrated, particularly in inner-city and peripheral areas, that there is often a mismatch between expenditure and need, and therefore a lack of territorial justice. It has also been shown that those areas with the greatest needs not only lack resources and are starved of public funding, but also wield insufficient political power to be able to influence the allocation of public funds.

Territorial Social Indicators

Social measures that are employed to monitor and assess spatial variations in the quality of life. A whole range of criteria may be used, relating to such things as income, health, nutrition, education, housing, social order, etc. Some researchers have suggested that the indicators fall into three groups:

i. those that relate to the actual level of territorial well-being;
ii. those that identify the specific deficiencies or needs of areas;
iii. those that relate to the effectiveness of alternative ways of meeting those needs.

Territorial Waters

The coastal waters over which a bordering state has jurisdiction. Under international law this was originally defined as a distance of 5 km. In 1958 the law of the sea convention extended this to 22.2 km (12 nautical miles), as well as clarifying the rights of states to the resources of the continental shelf to a distance of 370 km (200 nautical miles), now known as the exclusive economic zone.

Tertiary Sector

One of the four major sectors of the economy, comprising the distributive trades, i.e., retailing, wholesaling and transport. Other tertiary activities include personal services, the professions and public administration, so that service provision is also a significant aspect of the sector. The tertiary sector may be seen as providing a link between many primary and secondary activities and their final customers.

Thaw Lake

A shallow, rounded depression containing a circular or semi-circular pond, in a lowland periglacial region. Thaw lakes are very common features of Thermokarst, and result from the surface thawing of frozen ground (e.g., at points where the vegetation cover has been disrupted). This produces localized collapse, and the formation of an irregular depression that becomes occupied by meltwater. The pond expands laterally rather than vertically through further melting of the permafrost and undercutting of the surrounding vegetation mat. The pond then becomes progressively smoother and more circular in outline.

Figure 66. Spring thaw.

Source: Image by flickr.com.

Thermal Pollution

Pollution of the environment (mainly the atmosphere, rivers and oceans) resulting from the release of heat from human activities, e.g., The escape of industrial and domestic heat into the atmosphere (contributing to the urban heat island effect), or the discharge of hot water into rivers where it causes de-oxygenation, adversely affecting fish and other aquatic fauna.

Thermokarst

The formation in periglacial environments of a highly irregular ground surface, as a result of the thawing of masses of ground ice. Hummocks, pits and larger enclosed depressions may become so numerous that there is a crude resemblance to the karstic features of limestone (though the processes of formation are totally different). Among the main features of Thermokarst are alases (and alas valleys), pingos and thaw lakes. The development of Thermokarst landscapes results either from a warming of the climate or from a removal of the vegetation cover that insulates and protects ground ice bodies. Construction work can lead to very rapid Thermokarst formation. In a few years the ground surface subsides, and becomes so irregular and marshy as to be virtually impassable to traffic.

Thiessen Polygon

Frequently used as an alternative to quadrat analysis in the analysis of spatial distributions taking the form of point patterns. The polygons are created by drawing a straight line between each point and its immediate neighbors, and bisecting those with new lines drawn at right angles. The latter intersect to form polygons. It is assumed that each point dominates the area defined by its polygon. The technique may be used, for example, to calculate the average amount of precipitation received over the total area of a drainage basin on the basis of data collected at a small number of rain gauges located at different points in the basin. Polygons are constructed around each rain gauge site in the manner described above, and the area of each polygon is calculated and expressed as a fraction of the total drainage basin area. For each rain gauge site, the amount of precipitation received is multiplied by its fraction of the total area. The resultant values for all the sites are then summed to give an estimate of the mean precipitation total for the whole basin.

Third World

A term used, often rather loosely and along with a range of broadly synonymous terms (e.g., Developing world, the South, LEDCS), to denote relatively poor and under-developed countries located mainly in Africa, Asia and Latin America. The term originated after the Second World War and in the context of the Cold War between the First World (capitalist) and the Second World (socialist). It refers to a growing group of non-aligned countries, many of which have recently achieved political independence from colonial powers.

While the Third World undoubtedly embraces some of the poorest countries of the world (e.g., Ethiopia, Bangladesh), not all of the constituent countries can be regarded in this way. For example, some countries are significant producers of oil (e.g., Nigeria) and minerals (e.g., Brazil); some have quite well-developed industrial sectors (e.g., India). So, the Third World might be seen as comprising either what are popularly known as the LEDCS or a mix of developing countries (i.e., Those in the process of achieving as regards organization, social systems and independence) and undeveloped countries (i.e., Those that have yet to make significant strides as regards economic development, rising living standards and political independence). Because of this diversity, it is clear that there is no single course of action for dealing with the different economic, political and social challenges facing the Third World. The broad strategy for the future may, in many cases, lie in development, but how that development might take place, in what form and at what pace are questions that can only be answered in the light of the circumstances that prevail in the individual countries.

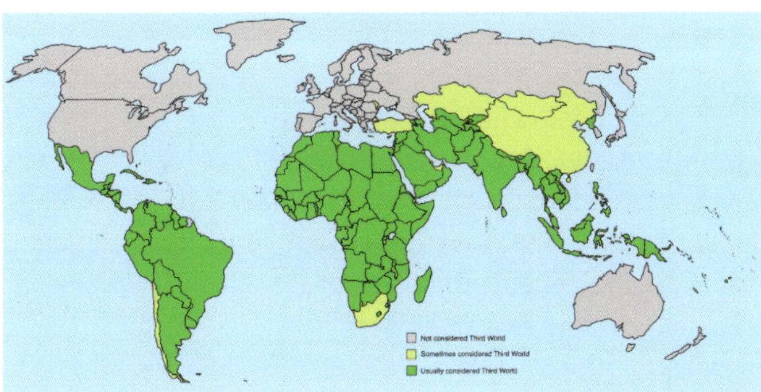

Figure 67. Map of third world countries.

Source: Image by Wikimedia.

Threshold

(i) A factor that complicates the self-regulation of systems by negative feedback, and thus the maintenance or restoration of states of equilibrium in open systems. When a threshold is crossed, rates of operation of natural processes may be rapidly accelerated, or irreversible changes are set in motion. A simple example is provided by the melting of glaciers early in the ablation season. At the end of the winter, glaciers are snow-covered (even in their lowermost parts), and owing to the high albedo of white snow, much solar radiation is reflected, and ablation is very slow. However, when the snow is eventually removed, and the darker and often dirt-stained ice surface (with a much lower albedo) is exposed, the rate of melting will be greatly increased.

(ii) In central place theory the term thresh-old refers to those conditions that any good or service requires for entry into a central place system. Before any good or service is offered for sale at a central place, all the costs involved in its production and provision must be covered by anticipated transactions. This minimum number of sales represents the particular threshold requirement of each good or service, while the threshold population is defined as the minimum number of people required to support any central place activity before it can be profitably operated. Although the term threshold is usually interpreted in this way as simply referring to the number of people or customers required, strictly speaking it should relate to that part of the expenditure of each person used to purchase each particular good or service.

Thrust Fault

A reversed fault at a very low angle. In major geological structures (such as nappes), the mass of rocks overlying the almost horizontal fault (also referred to as the thrust plane) may have ridden forwards a distance of many kilometres over the rocks beneath.

Thunderstorm

A storm characterized by thunder and lightning and heavy, even violent, precipitation, resulting from very rapidly rising air currents under conditions of absolute instability. Thunderstorms are associated with the formation of anvil clouds of great vertical extent, within which up currents may attain rates of 30 m s^{-1}, and extend as high as 6000–12,000 m. Rapid condensation leads to the formation of both water droplets, and HAIL and ice crystals. The release of latent heat is an additional factor promoting rapid uplift of air. Within the thundercloud, positive electrical charges are built up (i.e., by the break-up of large raindrops carried aloft by the updraughts). When these are discharged, either to areas of negative electrical charge within the cloud (associated with hail pellets) or to the negatively charged Earth, lightning results. One important feature of thunderstorms is the formation of downdraughts of cold air (to compensate for the rising currents), which in the later stages of the thunderstorm exceed the updraughts and lead to the degeneration of the 'thunderstorm cell.' These downdraughts carry down, often at a high velocity, the raindrops and hail from the upper part of the cloud, giving rise to violent squalls.

Tide

The twice-daily rise and fall in the level of the sea, resulting from the gravitational attraction on the Earth's oceans exerted by the Sun and Moon. The tidal range (the vertical interval between high and low tides) varies both from time to time and from place to place. Tidal currents are stream like flows of seawater into and out of bays and estuaries with the rising and falling of the tide producing respectively flood currents and ebb currents.

Figure 68. Waves tides sea.

Source: Image by Pixabay.com.

Till

A deposit laid down by a glacier or ice sheet on a land surface. Till is highly variable in character, depending on the precise manner of deposition, but it is generally highly mixed (with particle sizes ranging from clay to boulders) and poorly stratified. Till may result from the basal melting of debris-rich ice or from the surface melting of the ice. The terms till and moraine are sometimes regarded as synonymous; however, strictly speaking, till refers to the deposit itself, and moraine to the surface landforms (such as an elongated ridge) of the till. In being laid down directly by ice, the extent of till can, theoretically at least, be used to suggest the extent of ice flow.

Till-Fabric Analysis

The measurement and analysis of the constituents of glacial tills, with the object of reconstructing the past glacial history of an area. One simple method of till-fabric analysis is to measure the orientation and dip of the long axes ('a axes') of a sample of 50–100 elongated stones contained within a till. These are then plotted on a rose diagram, designed to show the preferred orientation (if any) of the stones. The assumption is usually made that, in lodgement tills (or ground moraines) formed by the slow bottom-melting of an icesheet, these stones (which have become aligned, within the ice, parallel to the direction of flow) retain their orientation and dip as deposition proceeds. Thus, till-fabric analysis allows the direction of ice movement to be reconstructed even from very old tills, provided that these have not been disturbed by localized slumping, periglacial freeze-thaw disturbance, or later ice advances.

Time-Distance Convergence

The reduction in the travel time between places and the decline in the importance of distance brought about by improvements in transport and communication. By reducing the friction of distance, the world is 'shrunk.' It is an important concept in understanding changing spatial distributions, spatial interactions and spatial relationships.

Timesharing

A relatively recent phenomenon in the contexts of recreation and tourism, which allows people to have a small stake in a holiday property (or second home). Developers construct blocks of flats, undertake chalet developments and subdivide large country houses into flatlets, and then sell each unit, fully furnished, on a weekly basis. Each customer buys the unit for a particular week or weeks either for a specified number of years or in perpetuity. The sale price of each week is determined by its timing in the year and therefore the weather expectations at that time. In the case of coastal resorts, high-summer weeks will be at a premium, while for winter ski resorts the guarantee of good snow conditions will be imperative. Similarly, those weeks coinciding with public holidays and school holidays will tend to be more expensive.

Time Lag

On a hydrograph, the time between peak precipitation and peak river discharge. It is a measure of how 'flashy' a river basin is, and how quickly it responds to precipitation.

TNC (Transnational Corporations)

Enterprises that, because of their size, their arena of operation and their merger of firms scattered in many countries are often known, in short, as the multinationals. Not only are they multi-plant enterprises, but they are also multiproduct. The general trend has been for a larger and larger share of economic activity at the regional, national and international scales to be performed by a relatively small number of extremely large business corporations. Direct investment by TNCs today is thought to account for more than one-fifth of total industrial output of the non-communist world. They penetrate nearly every country and, in almost all cases, they are increasing their share of gross domestic product; in some cases, they are responsible for more than one-third of total manufacturing output. TNCs are more prominent in some branches of economic activity than others. For example, they are dominant in the high-technology industries (e.g., IBM); they are also very conspicuous in motor vehicles (e.g., Ford), chemicals (e.g., ICI), mechanical and electrical engineering (e.g., GEC) and oil (e.g., Exxon). They are key players in economic globalization.

Tombolo

A sand or shingle bar, resulting from the extension of a spit by longshore drift or the migration of an offshore bar towards the coast, and linking an island with the mainland.

Topography

The description of the surface features of a place. Topography, strictly speaking, refers not only to physical features (in which sense it is used frequently though incorrectly), but also human features (settlements, communications, etc.). A MAP that shows these features in detail (e.g., the 1:50,000 and 1:25,000 Ordnance Survey maps of Great Britain) is known as a topographical map.

Topological Map

A map based on data that have been subjected to topological transformation. On the map, some basic aspects of the real world, like boundaries, the original number of locations and linkages, are faithfully reproduced, but distance and direction are subject to distortion. The transformation or distortion is undertaken to enhance communication of information and to eliminate irrelevant factors. As a result, the topological map assumes a diagrammatic quality. The map of the London Underground is a very famous example of a topological map. Although there is no scale and no accurate portrayal of direction, the map is effective in showing the individual stations and the different lines that make up the underground system. The topological map is widely used in network analysis. Sometimes referred to as a cartogram.

Toponymy

The study of place names, particularly of their derivative elements, "linguistic" origins and meaning. The evidence contained in place names about former aspects of the environment (physical conditions, patterns of settlement and colonization, cultural diffusion, etc.) has been an important source material in historical geography.

Tor

A rocky outcrop most commonly found on hilltops, varying in scale but usually less than 30 m in height, in which the vertical and horizontal joints are clearly exposed. Tors are characteristic features of granite landscapes, although they are not restricted to these. One theory of tor formation is based on the assumptions of variably jointed granite and an episode of deep chemical weathering under warm and humid climatic conditions. Weathering is intense where the joints are closely spaced but much less effective where the joints are more widely spaced. When, in due course, the regoliths is removed (by solifluction or fluvial action), the less weathered section of granite is left upstanding to form a tor.

Tornado

A counter-clockwise whirling storm, formed around a small cell of very low atmospheric pressure. Wind speeds are exceptionally high (sometimes in excess of 300 km hr^{-1}), causing serious structural damage to buildings along the narrow storm path. Tornadoes develop frequently in the Mississippi Basin, at the line of junction between warm, damp air from the Gulf of Mexico and cool air from the continental interior, mainly in spring and early summer when there is the additional factor of thermal heating of the ground surface. Tornadoes are usually very short-lived, lasting for only an hour or two, and are limited in size (usually only 100 m or less across). Their destructive effects are also very localized, though they can cut a swathe across a town.

Tourism

Leisure-time activity generally denned as involving at least an overnight stay away from home, thereby distinguishing it from recreation. Tourism frequently involves an important commercial dimension, with investment being made in hotels, motels, caravan sites, etc., in the provision of a diversity of tourist-oriented services, and in the improvement of access to locations favored by tourists. Thus tourism, through its generation of employment and economic wealth, can make a significant contribution to the local and regional economy. Indeed, tourism is currently the largest and most rapidly expanding industry in the world. Where tourists are attracted in large numbers from other countries, tourism can create invisible earnings and make a notable difference in the balance of payments of a country.

Figure 69. Tourism is the highly developing sector of business.

Source: Image by Pixabay.com.

Tower Karst

A type of tropical karst landscape, more 'advanced' in development than cockpit karst. Tower karst comprises steep-sided, isolated masses of limestone (mogotes), which can be regarded as the equivalent of the inselbergs of granitic terrains. The for gradually been undermined by basal weathering, leading to the formation of many individual residual masses, small in extent but of considerable height. These 'towers' rise up to 1000 m or more above a near-level plain, resulting from prolonged solution at the level of the permanent water table.

Town

A compact settlement, larger than a village, with a community pursuing an urban way of life.

Townscape

In essence, this is the urban equivalent of landscape and comprises the visible forms of the built-up area, particularly street plan and layout, architectural styles, land and building use.

Toxic Waste

Chemical waste that is poisonous to people and wildlife. Until recently, the most common way of to get rid of this was to dump it in landfill sites. The risks of ground water pollution is now such that it is increasingly common for such waste to be disposed of by incineration. However, the pollution risks to the atmosphere do seem to be considerable. In theory, the movement of hazardous waste across national borders is prohibited.

Trace Element

An element that occurs in the soil in minute quantities, but that is of vital importance to growing plants. However, if the trace elements are too abundant, they can actually make the soil poisonous. Valuable trace elements include manganese, boron, molybdenum, copper, cobalt and lead; these are often provided by the weathering of basic igneous rocks.

Trade

A flow of commodities and services from producers and providers to consumers. The generation of trade may be explained in terms of either comparative advantage or of exchange relationships within and between different modes of production. Trade is a vital aspect of economic development as well as a prime example of spatial interaction.

Trade

Group of countries drawn together by trade agreements that promote free trade between them. At the same time, the group as a whole is protected from other countries by tariff walls. The EU and NAFTA are two such blocs that account for a large proportion of global trade.

Trade Friction

A term used to describe the tensions that tend to surface when there is imbalanced trade between two trading partners, as has occurred between Japan and the USA. The trading balance between these two countries shows a huge surplus in favor of Japan. As a result, the USA has harboured feelings of resentment; it has also tried to explain its lack of success by accusing Japan of indulging in unfair trading practices. This, in turn, has led to threats of retaliation in the form of protectionism. For its part, Japan has felt offended by these accusations and threats.

Trade Gap

This is the shortage of the exports as compared to the imports.

Trading Estate

This is a planned industrial estate which is basically designed by local or the national government, primarily to diversify employment and the opportunities in area of high unemployment or of the unequal industrial structure (as where there is undue dependence on the very main and the basic industry for example, as it is the England). Also, there is a constructed around the outlines of the towns and the cities so that they can accommodate the industry which are relocated from the inner urban areas. This also gives the facility and also includes the services of the standard design's factories and the centralized information, publicity and the administrative services.

Traffic Principle

One of the principles among the others underlying the Christaller's central place theory and the governing the spatial arrangement of the central places which is related to their hinterland. The traffic principle which sometimes also is referred to as the transport principle. Which applies in the central places are located so that the lower order of the centers which lie along the straight-line paths between the higher order centers. These arrangements usually have k value of the 4 which shows the most efficient and the rational arrangement from the point of the center view of the network of the road which is required to link the system of the central places together.

Transfer Cost

The total sum of the money that should be offered to attract the supply and the factor off the production like the labor capital, etc. This is also away from the alternative uses. This is the total sum of the necessary supply price of that factor which is the new use. The transferability also refers to the transport cost which have the characteristics of the different commodities and also to the handling various different types of the goods are now more sensitive to distance others, while some of the goods are more easily handled than the others. The high transferability would also indicate that the commodity has a very high value which is in relation to its transport costs and that it is also readily transportable.

Transfluence

It is formed when the outflow of a glacier (or ice sheet) is impeded, its thickness consequently increases, and ultimately a new escape route (the breach) is exploited.... The many breaches through the western Highlands of Scotland are due to the accumulation of transfluent ice east of the main watershed.

Transshipment

A transshipment is when cargo or a container is moved from one vessel to another while in transit to its final destination. An excellent example of a transshipment would be the transport from Durban to Manila because there is no direct connection between the two ports.

Transhumance

Transhumance is a type of pastoralism or nomadism, a seasonal movement of livestock between fixed summer and winter pastures. In montane regions (*vertical transhumance*), it implies movement between higher pastures in summer and lower valleys in winter. Herders have a permanent home, typically in valleys. Generally, only the herds travel, with a certain number of people necessary to tend them, while the main population stays at the base. In contrast, *horizontal transhumance* is more susceptible to being disrupted by climatic, economic, or political change.

Transit Trade

Freight traffic which goes through from one country to another over a third. For instance, an extensive volume of freight set foot in The Netherlands and goes up the Rhine to Germany as well as Switzerland, whereas Luxembourg obtains substantial revenue from transports rail shipment across its territory.

Translocation

The transmit of material enclosed in a SOIL through solution or in suspension resulting from one horizon to the next.

Transmigration

A government conducted scheme in Indonesia to inspire people from the overpopulated, productive islands of Java and Bali to relocate on the distant, lower density of population islands of Sumatra, Kalimantan, and Irian Jaya. It is the world's immense effort to transfer people. The original program was to move approximately 68 million people between 1984 and 2008.

Transpiration

The disappearance of water vapour from a plant via the minute pores *(stomata)* that shield the leaf surface. The proportion of water transpired will be eventually affected by the structure of the plant (that might be adjusted to limit water losses, mainly in arid environments), and factors such as temperature, humidity, and wind speed. An enormous tree (e.g., a fully grown oak in England during summer season) might transpire various hundreds of litres of water every day; a maize plant, on the other hand, will transpire merely 2–3 litres per day.

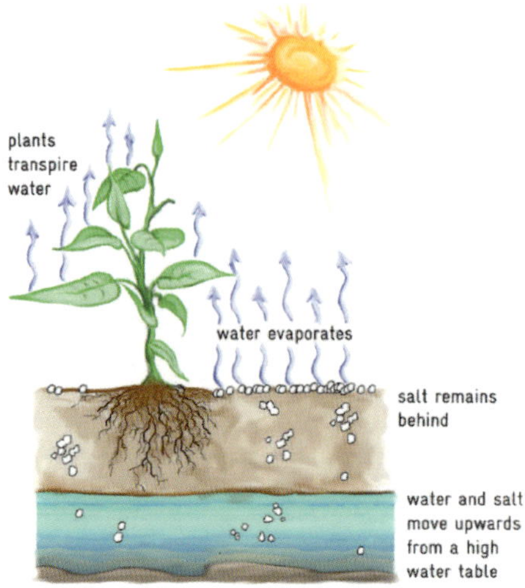

Figure 70. Schematic representation of transpiration.

Source: Image by flickr.com.

Transport

In the geomorphological aspect, the movement of sediment through an agent for example, running water, wind, glaciers, icesheets, crashing waves as well as the tidal currents, (ii) transferring people, goods, and services from one location to another through several modes, most likely, road, rail, and air.

Transport Cost

The aggregate cost of transferring a good, mainly proportional to weight or volume and to distance it is carried. Speaking on a serious note, the cost must also take into consideration the expenses incurred in packaging, insurance and of trading with the paperwork that generally go along the movement of commodities (e.g., completing customs forms, invoices, etc.). Transport costs precisely portray a crucial production cost. On the other hand, it may be argued that technological progress in transport (pipelines, bulk carriers, etc.) Have deducted the general importance of transportation costs, the increasing costs of fuel might be viewed as working on the contrary direction.

Transport Geography

The study of geographical dimensions of transport that in the recent years include such varied themes as spatial strand of transport (networks, terminals, flows of commodities and passengers), the role played by transport in the form of an agent of spatial transitions (mainly in the context of development) and the influence of transport on accessibility and mobility in particular domains and with respect to various social groups.

Transport Network

The transport passages (roads, railways, canals, etc.) Associating a set of nodes (i.e., junctions and terminals); the connections between beginnings and destinations.

Transport Network Development Model

This model demonstrates six consecutive phases in the development of a transport network. It begins with a sequence of small ports dispersed along a coastline, each being abandoned and containing its own small hinterland. In the next phase, some of those ports expand more rapidly as compared to the rest and evolve longer lines of transport and communication on the inside, while in the succeeding stages intermediate centers develop throughout these lines. In the course of next two stages, there are rising degrees of fusing and correlation between the two lines, which takes eventually in the final stage to the development of high-priority routes giving direct association between the most significant.

Tree Line

The 'line' highlighting the limits of the growth of a tree. The tree line mentions to both the as the Alps, where the transformations from the bottom forested slopes towards the upland grassy meadows is at times quite immediate) and the latitudinal limits (as on the northern fringes of the boreal forest, where the tree line is more of a wide transitional zone in between the coniferous woodland and the dwarf shrubs of the TUNDRA). In the USA the tree line is referred to as the *timberline.*

Trellised Drainage

A pattern of drainage is sometimes linked with scarp-and-vale landscapes and specified by right-angled stream intersections. The important consequent streams goes along the bands of resistant and unresistant rocks, following the pathway of dip. The frail strata are eroded into strike vales by subsequent streams, which often accounts for the capture and disturbances of consequents. Trellised drainage is therefore fixed well to the geological framework.

Triangular Diagram

The arrangement of three related or connected dimensions of some attributes or item on triangular graph paper, imputing a higher value for each strand to each apex of the triangle; for instance, three facets of climate (that is mainly pressure, temperature, humidity), population age (young, middle-aged, old), sediment (sand, silt, clay), sectoral employment (primary, secondary, tertiary). Often known as a *ternary graph.*

Trickle Down

The expression used by Hirshman (1958) to signify the spread of evolution from the core to the periphery.

Trophic Level

A magnitude of production of food or consumption within the framework of food chain or ecosystem. Nearly all food chains constitute three important trophic levels: *primary producers* (autotrophs), *primary consumers* (herbivores), and *secondary consumers* (carnivores).

Tropopause

The intersection between the troposphere below and the stratosphere upwards.

Tropophyte

A plant that is modified to hold out against seasonal periods of cold or drought. Tropophytes therefore act as xerophytes throughout the part of the year (when soil moisture is not easily available and/or temperatures are way too low for) and hydrophytes in course of the season of growth. Predominant examples of tropophytes are the oak, beech, and birch of temperate deciduous woodlands, having their seasonal behaviour of summer growth and winter 'rest' (acquired by the detachment of leaves).

Troposphere

The underneath section of the atmosphere, lying at the bottom of the tropopause. The troposphere differs in depth, ranging from an average of 18 km above the Equator to just 6 km exceeding the poles. Nearly all-weather phenomena take place within the boundaries of troposphere.

Trough of Low Pressure

An area of proportionately low atmospheric pressure depicted on a synoptic chart by means of an addition of the isobars present in one direction from the focal point of a depression. A trough of low pressure is most commonly linked with the directionless showery weather.

Trough End

The rapid head of a glaciated mountain valley *(glacial trough),* frequently associated with near-vertical rock faces and breath-taking waterfalls (e.g., as at the head of the Engstligen valley in the Bernese Oberland, Switzerland). Trough ends are considered to be the outcome from massive powers of 'over deepening' influenced by a large valley glacier that has been constructed by the consolidating of a number of smaller, much less robust glaciers by emerging of high-level cirques beyond the valley head.

Truncated Spur

A spur or a stimulant, previously projecting into a preglacial river valley that has been disintegrated in its lower segment by a hefty valley glacier, which would have been powerless to go along the same sinuous path as the preglacial river. Truncated spurs are very usual and characteristic landforms of glacial troughs for instance the Lauterbrunnen valley, Switzerland.

Tsunami

A very immense seismic sea wave, which rises due to an earthquake shock which generates on the ocean floor. Tsunamis travel for substantial distances across the sea in the form of a long, low waves (the wave length might be over 100 km, the wave height as short as 1 m, and the forward velocity as much as 600 km hr^{-1}). As the tsunami reaches the shore, the wave height rises distinctly, and often go beyond 15 m; it is therefore efficient of resulting into an immense demolition to coastal settlements and extreme loss of life. For instance, the great Krakatoa eruption which occurred in 1883, connected with seismic disturbances, gave rise to tsunamis that drowned and took away the life of 36,000 people in coastal villages of Java and Sumatra.

Figure 71. Representing damages caused by tsunami.

Source: Image by pixnio.com.

Tufa

A deposit caused from the precipitation of calcium carbonate, or *calcite,* from calcium bicarbonate diffused in water. The process of precipitation, in which carbon dioxide is released, is most efficient and useful when there is a rise in temperature or a decline in pressure. Thus, tufa is commonly established at points where springs appear from passages below the grounds in limestone or chalk terrains. Inside limestone caverns, characteristics such as stalactites and stalagmites are constructed resulting from tuffaceous deposits.

Tundra

The zone lying between the latitudinal limits of growth of a tree and the polar ice. (In actuality, the tundra is a trait of the Northern Hemisphere only.) Winters are harsh and deadly, with temperatures that drips down as low as –30°C at the northern limits. The growing season is short and cool with the mean temperature of the warmest month hardly rising beyond 10°C.

Tundra is connected with the wide-scale development of permafrost, which limits the root penetration and leads to water-logging of the soil in most of the regions during the season of summer. Soils are predominately skeletal, owing to their pre-eminence of freeze-thaw weathering, and the vegetation constitutes of mosses and lichens, low woody plants and deciduous dwarf shrubs which are up to 2 m in height (willow, alder, birch) across the southern, less critical extremity of the zone.

Turbulent Flow

A kind of flow most commonly linked with rivers in which there is a diversification of secondary movements (eddies) transient on to the significant forward movement of water. Turbulent flow in rivers is essential for balancing the suspended load, which is persistently being elevated by the topmost eddies. Turbulent flow also takes place within the lower atmosphere provided the conditions when airflow is turned upside down by surface interferences for instance buildings and areas of irregular relief.

Twilight Area

Generally used with reference to inner city parts in which there is a common collapse in the situations of the urban fabric and in recognized status. The run-down character is the outcome from causes such as: the normal obsoleteness of buildings with the course of time; the failure to shape well established investment in their preservation and use; the rising obtrusion of non-residential uses into earlier residential areas; the intensifying gathering of poorer households in the low-quality housing, etc. Twilight regions are often a characteristic of the zone of transformations.

Typhoon

A tropical whirling storm, characterized by winds of tremendous velocity and torrential rainfall, taking place in the China seas and throughout the western brinks of the Pacific Ocean. Typhoons are referred to as a local type of tropical cyclone or hurricane.

U

Ubac

A hillslope which, in the Northern Hemisphere, having their sides at northwards or northeastwards, and therefore obtains the less amount of sunshine as well as warmth.

Ubiquitous Material

As stated in the weber's theory of industrial location, this term is considered to be any material that is accessible everywhere (for instance, water) and thus does not apply a pull of industrial location.

Umland

Refers to the German term which stands for the hinterland, market area or sphere of influence of a central place.

Unconformity

A break in the progression of a series of rocks. The older rocks below the unconformity are frequently folded or tilted, while the younger rocks placed above are most commonly of simpler structure (even can be horizontal). The unconformity demonstrates a period of erosion influencing the surface of the older rocks, leading in some instances to the construction of a peneplain. If this surface is then, for instance, impacted by a marine transgression, then the younger sedimentary rocks will be settled or in other words deposited on it.

Figure 72. Angular unconformity.

Source: Image by flickr.com.

Undercliff

A heap of slumped material at the foot of a sea-cliff that is being impacted by an extensive mass movements. The presence of an undercliff (commonly in areas of unclear sands and clays, or in permeable chalk overspreading impermeable clay) denotes that subaerial cliff downturn is at the minimum as effectual as marine erosion in the severe perception of the term.

Undercutting

Lateral explosion at the bottom of a slope – for instance, the undermining of a stream bank by a turning stream, the lowly erosion of a valley slope done though a sidewards shifting river, or the construction of a wave-cut mark at the bottom of a sea-cliff.

Underdevelopment

The economic prosperity of a country or region in which, speaking on a wide-scale, there is chance for an entire exploitation or consumption of resources as compared to what is being obtained in the recent years. Therefore, underdevelopment might be said to come out ahead in any area that has great potential probabilities for the utilization of more capital, more labor and more resources that are accessible in order to support its current population at an inflated level of living or, if its per capita income is so far reasonably high, to back up an even large amount of population at the similar standards of living conditions.

Therefore, underdevelopment from an economic point of view refers to a relative state that prevails at a scale of structural scales. It unquestionably exists in the third world, yet uniformly it may be said to exist in areas and regions of what would be represented and identified as developed countries (e.g., in the periphery areas of such countries). Few of them would contend that underdevelopment also has significant political and social aspects.

Underground Stream

A stream passing through an underground passage, most commonly withing limestone. Underground stream usually initiates as surface flows on impermeable rocks adjoining to the limestone outcrop. They set their foot in the limestone by the mode of swallow holes and later on re-emerge, at times at the head of steep-sided ravines, in the form of resurgences.

Underpopulation

This resources and development could hold a great amount of population in the absence of any lowering of the level of living or where a population is insufficient to evolve and employ its resources effectively. The previous situation takes place in areas of Australia and New Zealand, containing the situation of extensive agriculture which is capable of supporting quite a high standard of living; areas of less population or depopulation in Western Europe would also come under this specific category. Examples of the latter situation happens where small technical levels exists, as surrounded by pastoral nomads in semi-arid parts, or in regions of pioneer settlement, as in the distant sections of Australia and Siberia.

Unemployment Rate

The number of people at the age of employment who are willing to work but cannot to find jobs evaluated as a percentage of all those of working age available for job opportunities at a given time period.

Uniclinal Structure

A geological structure consisting a sequence of sedimentary layer dipping approximately uniformly in a specific direction. Uniclinal framework comprising staggered resistant and unresistant beds gives absolute conditions for the progression of *scarp-and-vale landscapes.*

Uniformitarianism

An elementary geo-logical principle, which highlights that Earth processes perceived today are intrinsically similar to those that are managed in the earlier times. Stating in brief, rocks are being constructed today in the same ways as they were formed hundreds of millions of years ago. Although, uniformitarianism must not be expounded in too simple a fashion. It does not signify that there have never been transitions in the proportions and frequency of Earth processes. Indeed, during the recent times, many such processes are being boosted up by various human activities.

UNO (United Nations Organization)

This initially came into existence when the Second World War in 1945 came to end, with the purpose of maintaining international peace, safety, and security, and of instituting the correct sort of political, economic, and social conditions for the apprehension of that aim. There are now around 160 endorsers to the United Nations Charter. Within the UNO a scope of specialist agencies has been formed, the most essential from a geographical perspective include: the FAO (Food and Agriculture Organization), GATT (the General Agreement on Tariffs and Trade, now the WTO), the IMF (International Monetary Fund), UNESCO (the *United Nations Education, Scientific and Cultural Organization),* the WHO (World Health Organization), the WMO *(World Meteorological Organization)* as well as the world bank.

Unstratified

Refers to a term implemented to deposits that are not distinctly layered.

Urban

However, extensively used in the explanation of places, populations and settlements, it is hard to give a simple, crystal-clear explanation far away associating to, feature of, a town or city. This is partially explained through the fact that the urban circumstances connect to a number of various, but inter-dependent dimensions, of which four might be considered as being specifically important,

i. Urban settlements substantially have greater populations as compared to the rural villages, though the size approach will differ from country to country,
ii. Urban places are characterized by higher population densities formed by the contiguous concentration of activities and buildings,
iii. In economic aspects, urban settlements and places are regarded nearly all entirely with non-agricultural activities and, in many examples, with the supply of goods and services for hinterlands,
iv. The people who reside in urban places are thought to take part in an idiosyncratic way of life.

The degree and accurate mix of these features vary from place to place (e.g., third world cities display features not experienced in MEDC cities) and occasionally (that is urban Britain in recent times shows quite different attributes as compared to those that pointed out the urban growth of 100years ago).

Figure 73. Urban street cityscape.

Source: Image by Pixabay.com.

Urban Blight

A decline in the physical condition of a metropolitan city or a reduction in its general standing. It most often takes place when there is some doubt or unpredictability about the near or distant future. Rumors about possible redesign programs or obligatory purchase, for instance, might fuel these kinds of uncertainty, which in its turn prevents property owners from establishing proper investment in the development of buildings. In the case of residential areas, urban blight may have more of a social aspect and incorporate some depletion in social status or appeal.

This might, for instance, be caused by the invasion of non-residential activities, by the accumulation of a minority group or by the transformation of single-family accommodations into the structure of multi-family use. Well -formed families are progressively induced to shift elsewhere, only to be replaced by residents or landlords who, usually, are either less able or less willing to invest in preservation of property. Therefore, obsoleteness and the collapse of fabric set in and as a result the blight proliferates.

Urban Climate

The distinguished climate connected with a wide-scale urban area. One prime feature of urban climates is the taking place of importantly higher temperatures as compared to in the rural surroundings. Other impacts of built-up areas are: lower atmospheric humidity (owing to rapid run-off of rainfall over impermeable surfaces and decreased evapo-transpiration); a rise in the environmental lapse rate (resulting from the appearance of rapidly heated road surfaces and roofs of buildings), construction of convectional rainfall along with thunderstorms more credibly; local transitions in the pattern and speed of winds, as the air is channeled between tall buildings and throughout narrow streets. In cities where there are clear and visible concentrations of impurities in the air radiation fog, or even smog, is expected to form under conditions of atmospheric stability and inversion of temperature.

Urban Conservation

The general purpose which it serves is at retaining or protecting the conventional appearance of a town's fabric, that is to balancing, renovating, and improving those sections that have character and demonstrate good instances of past achievements in urban framework. At the same time, although, urban conservation will at times incorporate looking for new functions for old buildings, adapting old structures to implement in new uses, together with allowing a specific amount of development, given the fact that it is in harmony with what already is available. In this context, it is a form of urban renewal, and it is essential to focus on the factor that good conservation does not look for conservation in a 'fossilizing' aspect.

Urban Fallow

Urban land that is in recent times unused, plausibly because it is in the procedure of redevelopment or due to the fact that it is the victim of inner-city decline.

Urban Field

Refers to a term which is previously used by British geographers when mentioning to the area located around, and efficiently associates to, a town or city. Other terms are now more extensively implemented, for instance. Hinterland, market area, sphere of influence, *tributary area.*

Urban Geography

That branch of geography which focuses on the location and structural arrangement of towns and cities, looking for to illustrate and describe both the distribution of urban places, and the resemblance as well as the contrasts (in economic and social dimensions) that prevail between them. It is also associated with the internal dispositions of towns and cities, recognizing both the patterns (mainly of land use and social areas) and the processes that have shaped those patterns with the duration of time. A final sphere of interest lies in those modern issues that possess a spatial aspect, as for instance inner-city decline, housing, and traffic obstruction.

Urban Hierarchy

The upright classification of towns and cities as stated to a single variable, for example, the size of population, extent of built-up area or central place status. If each urban settlement enclosed in a provided country or region is organized on the basis of a dispersion diagram as claimed by the value it reports for the selected variable, the formation of plotted values may disclose a magnitude of either even arranging or clustering.

The situation mentioned earlier would be explained as denoting the existence of an *urban continuum* (with each urban center inhabiting an exceptional position throughout the variable). Whereas in the latter situation, each cluster might be held as a clear-cut class or order enclosed by the urban hierarchy; the lower the mean value of each cluster, the lower its status or status in the hierarchy. The disposition of regional as well as national urban systems may be in a way that the urban hierarchy is frequently viewed in the terms of a pyramidal structure, so that the lower the standard the greater will be the number of prototypical arrangements (meaning to say, the larger the cluster on the diffusion diagram). Simply put, in most urban social scale, it will be found that there will be more towns as compared to the cities, and more cities in contrast with the regional or provincial capitals.

Urban Land-Value Surface

Structural differences in rents or land values inside the built-up area, established by the bidding process in what is generally conjectured a free-market situation (refer to BID-rent theory). It is generated by spatial dissimilarities in the degree of demand and the locational attributes of individual plots. The urban land-value surface is assumed to exert a very hefty impact above the spatial structure of towns and cities, mainly the arrangement of uses of land and various social groups.

Urban Managers

Bureaucrats (that is planners, housing and welfare officers, or councillors) who, together with gatekeepers, apply substantial influence over the employment of limited resources when it comes to urban areas, as between various groups and between various localities. Their regulation is most closely perceived in relation to public goods and services, specifically with consideration to such matters as access to housing (remarkably, the council housing), and the equipping and giving of social and welfare services (like schools, clinics, community halls, etc.) In disparate sections of the city.

Urban Mesh

The network of towns and cities found in a particular area, perceived especially in context of spacing, structural pattern, functional connections, and connectivity.

Urban Morphology

The examination of the built fabric of towns and cities (from its layout, form, to functions, etc.) And of the methods in which this fabric has developed with the time.

Urban Planning

The procedure of organizing transformations in order to obtain particular purposes and aim in respect to the urban system. Planning activity may anticipate a number of different perspectives, as well as function at different spatial domains. For instance, much urban planning is beyond any question is regarded with the improvement of urban issues, such as making an attempt to correct imbalances prevailing in the urban system (e.g., devising programs of decentralization to convey growth from primate cities to lower orders of urban center in less benefited regions) or to ameliorate living conditions within individual towns and cities (by instituting housing programs, schemes and traffic schemes, and catering to the basic needs of people by providing better social and welfare services).

In contrast, urban planning can be much more progressive in the perception of projecting recent trends, foreseeing problems which is most likely to emerge and then implementing appropriate planning programs in order to reduce those problems and increase the benefits. Such a planning exercise might incorporate projecting the directions of inner-city decline and then putting together a proper scheme of action. The program might not certainly seek to hold against the decline, rather it might aim to deliberately phase the confiscation of people and jobs, and the curtailment of services.

By doing so, and by making certain their effective relocation and provisions elsewhere, the influence of the decline might be reduced. Finally, it is important to mention that the conventional segregation of town and country planning has given a path in recent years to a much broader approach for planning the issues, for it has been identified that the solution of urban problems will most of the times involve rural regions and vice versa.

Figure 74. A model city architecture.

Source: Image by Pixabay.com.

Urban Renewal

The modernization and reclamation of obsolescent urban areas by medium of either advancement (e.g., by appointing modern provisions in old dwellings, by road widening, etc.) Or redesigning (that is destructing all existing framework and beginning afresh). In the initial post war period, urban renewal programs tackled in British cities inclined to focus on redevelopment (no doubt encouraged by the devastating of bomb damage suffered in the course of the Second World War), but afterwards apprehension of its social costs (mainly where residential areas are concerned) giving rise to a shift of emphasis towards amelioration both of individual framework and also of the urban environment.

Urban Sprawl

A wide-scale unplanned, trailing, and low-density structure of urban or sub-urban development taking place around the boundaries of a town or city, specifically throughout radial routeways and at times leading to the amalgamation of once-separate settlements. Sprawl was a feature of much suburban growth in Britain in the time of the inter-war period when, in the absence of the strict planning regulations that now apply, considerable amounts of farmland were sold off to speculative builders, these investments were risky. In the USA regions of such development are often known as *slurbia.*

Urban System

The network of urban arrangements found in a specific area, but with each Urban settlement and its reliability on hinterland viewed as constituting a different, urban-centerd area. A significant feature of the urban system is their dependence on each other of its constituent settlements. No one town or city is entirely self-sufficient instead it depends to different degrees, on goods or services generated and provided elsewhere, that is by other urban centers.

The multiplex functioning of the urban system depends severely upon the growth of an associating transport network, for it is through this that the crucial movement of people, goods, capital, etc. Occurs, thereby facilitating exchange of communication between urban settlements, and between urban settlements and their hinterlands. In addition to this, an essential factor to this integration is the development of effective communications systems.

Urban Village

A village that has become swamped by the built-up area of a constantly growing city, but which still keeps hold of the part of its original character, its roots and identity (e.g., Fulham, Hampstead, and Islington in London). A residential district within a town or city and in which the people living there share a powerful sense of community, unity, and local attachment. In many examples, this aspect of identity stems from the grouping of households with same cultural attributes or sharing a common national removal. Furthermore, it is fortified by the provision of specialist services (shops, places of worship, clubs) in order to meet the specific needs of the group.

Urban villages are well represented in the inner areas of American cities, because of the concentration of ethnic immigrant groups in particular locations. The existence as well as social connection of such urban villages tend to deny the general assertion that the urban way of life (urbanism) is distinguished by obscurity along with non-discriminations.

Urbanism

The way of living, ethics and attitudes that distinguish people who reside in towns and cities. The question is progressively coming to light as to the magnitude to which urban and rural populations really vary in terms of these three mentioned bases. The shift of opinion is that, both in developed countries as well as the third world, nations the variations are obscure and are dimmed, if not they are inconsequential, distinctly since urban-based mass media offers to further encourage urbanism in country districts, well far off the limits of the built-uparea.

Urbanization

The procedure of becoming urban; a synthesis process of transformation influencing both people and places. Its main aspects are: a constant concentration of people and activities in towns and cities, through increasing the general domain of urban settlement; a transition in the economy of a country or region, whereby non-agricultural functions become supreme; a transition in the spatial attribute of populations (e.g., lower birth rates, higher degree of death rates, positive migration balances); an advancement of urbanism far away the built-up areas of towns and cities, thereby prompting *rural dilution* the imparting or diffusion of transformation (like economic, social, technological) down the urban social scale and into rural regions. Urbanization does not always necessarily take the same shape, nor does it advance at the same rate in all directions. In MEDCS, urbanization has now extended the phase of being instead more dispersed, by means of the proliferation of towns and cities, moreover by decentralization and through the advancement of urbanism beyond the built-up area. In LEDCS, urbanization tends to be more concentrated, having large rural-to-urban migration capacity converging on a restricted number of big cities.

Urbanization Curve

A model that may be used to chart the magnitude and advancement of urbanization in a given country or area. The curve takes the shape of an attenuated and often be subdivided further into three segments or stages:

i. The *initial stage:* during this phase the rate of urbanization is tremendously slow and only a small degree (less than about 25%) of the population live in urban settlements. In the course of this phase, the conventional society keeps going, with the emphasis placed on the rural sector of the economy and with an immensely scattered rural population.

ii. The *acceleration stage:* during this stage there is an intense redistribution of population brought about by enormous rural-to-urban migration. Population therefore becomes sequentially concentrated in a proliferation of towns and cities. This demographic transformation is greatly prompted by a basic restructuring of the economy, in which the secondary sector and the tertiary sector become superior. During this time, the urban constituent in the total population expands to between 60% and 70%.

iii. The *mature stage:* the curve starts to level off as the rate of expansion in the urban population begins to decline back to match the entire increase in population. In addition to this, the sectoral drift occurs in the economy, with the tertiary and quaternary sectors becoming more and more significant.

The onset during the current years, in some greatly urbanized countries of decentralization and counter urbanization might contend that a possible fourth stage is about to be introduced, in which there is an expectation that the urbanization curve might later on begin to fall away. For the moment, although, the losses seem to be hitting only the greater cities and to be matched by recompensing gains in those urban communities within the small to medium size domain.

The urbanization curve is a greatly generalized depiction of the pattern of change. At any point in time, various countries will get hold of different positions along that curve, whereas, over time, countries will advance at different rates. Therefore, for those countries going through particularly sudden urbanization, the curve will be very flattened along the frame of the time, while for those making slow advancement, the curve will be even more weakened.

U-Shaped Valley (Glacial Trough)

A 'typical' congealed upland valley, having sharp or near-vertical walls and a comparatively flat floor (which is often further focussed by postglacial alluvial deposits). The U-shaped valley is, in effect, the foregoing 'channel' of a valley glacier. It has been put forwards that the cross-profiles of glaciated valleys often estimated to mathematical curves (for instance, catenary curves); although, in reality profiles are immensely variable, mainly in the UK where 10,000 years' worth of postglacial processes are remarkable and have left their prints.

Figure 75. U-shaped valley.

Source: Image by Wikimedia.

Uvala

A circumscribed depression in limestone country, most commonly irregular in outline as well as perimeters and is a consequence from the composition of a various number of dolines.

V

Vadose Water

Vadose water is subsurface water between the land surface and the saturated zone below the water table. As a result, the vadose zone exists above the phreatic zone.

Valentin's Coastal Classification

There are three types of coastlines: advancing coasts, where the dominating land is either in deposition or uplifted. Spits, bars, elevated beaches, dunes, and salt marshes are typical coastal characteristics, as are receding coastlines when erosion or submergence of the land is dominating. Stacks, fjords, rias, and wave-cut platforms are some of the key attributes. Along with intermediate coastlines, which are situated in a stationary position, in which erosion equals uplifting or deposition equals submergence.

Value-Added

The income generated by sales of commodity and the cost of production, the gap between the two is known as value-added. In brief, it is the value that the manufacturing process adds to its factors of production. Value-added is used as a foundation for taxes in the European Union. Furthermore, the tax is usually a predetermined percentage of the price paid for a certain commodity or service rather than being imposed on the difference between the sale price and the manufacturing costs.

Value Judgment

Decisions made by an individual, which are guided by their own views, prejudices, beliefs, and other patterns of thinking.

Van't Hoff's Rule

The rate of growth of chemical reactions with respect to increasing temperature is indicated by the Van't Hoff's law. In general, the increase is 2–3 times for every 10°C rise. The crucial functioning of the process of chemical weathering in tropical landscapes can be comprehended via Van't Hoff's law.

Variable

Anything or a phenomenon that can take on a variety of different values. Values that do not have any clear demarcation or breaks between them are known as continuous variables. Variables of this type include length, weight, temperature, and time, where any value within a specified range can be considered. A discrete or discontinuous variable can only be measured in whole numbers or integers, such as the number of children in a household or the number of goals scored in a football game. Also, refer to; independent variable and dependent variable.

Variable Cost Analysis

An approach, to the analysis of industrial location that focuses on the costs (e.g., procurement costs, production costs) that vary spatially. A composite cost surface is generated based on variable cost input, and it is possible to identify the least-cost places on this surface.

Variable Costs

Firstly, a cost analysis can be described as those costs that vary with production scale; yet, efficiencies may break the simple mathematical relationship between production costs and production volume. Fixed costs, whereas, in geographical studies, variable cost is used for referring to costs that are prone to spatial fluctuations; location choices are impacted significantly due to these variations.

Variable Revenue Analysis

It is a method based on economic location research that focuses on the consumer side of the industrial production location dilemma. Instead of the cost side, encompasses variables such as cost analysis and market area analysis.

Varignon Frame

A mechanical model that may be used to identify the site of lowest transport costs when applying Weber's theory of industrial location.

The geolocation polygon is simulated by the model using the correct sized weights and pulleys linked by wires. The weights reflect the intensity of the attraction force at each polygon corner, whereas wire lengths are proportional to distance. The point of balance at which the linked wires come to rest is therefore believed to be the most cost-effective position (i.e., The optimal location).

Varve

A varve is a distinct band of sediment deposited on the floor of glacial lakes. They are typically only a few mm thick. Varves mostly comprise either coarser sediment (e.g., sand, silt) or finer sediments (clay). The coarser sediment is deposited during summers owing to the abundance of melt water from glaciers. The streams formed by the melt water can carry heavier and coarser sediments to the lakes.

Since there is no flow of melt water during the winters, the heavier sediments like sand and silt do not get carried to the lakes. However, during this period, the finer sediments like clay start to form a varve on top of the last summer's coarse-sediment varve. Each pair of these varves represents a year – comprising of a summer season and a winter season.

This information is useful in determining the age of a lake. One can count the number of these pairs of alternating varves and determine the age. We can also get information about climate change by checking the thickness of varves and hence the duration of the summer/winter seasons.

Method for examining moving phenomena that change direction and velocity (e.g., The tide or wind). A wind vector may be generated by accumulating the frequency of wind-force recordings from eight distinct directions.

Vegetation (Plant) Succession

All vegetation type plant communities experience a series of alterations from its establishment stage until the 'ultimate' vegetation cover (climatic peak vegetation) is reached, this process of progressive growth of vegetation type plants is known as vegetation (plant) succession. When a bare surface forms (e.g., as a result of glaciers, volcanic activity, or coastal sedimentation), it is initially inhospitable to plant colonization because it lacks soil cover and plant nutrients and may be excessively wet or dry.

As a result, the pioneer plant cover will struggle to establish itself and will be made up of tiny plants (mosses, lichens, and herbs). As the soil begins to form, a sequence of plant communities (known as serai stages) will emerge. Over time, serai communities will become more complex and comprise larger plants (there is a natural succession: grasses-shrubs-trees), with each community contributing to killing those which came before it. As humus from the developing plant cover is delivered in increasing quantities, and weathering contributes to soil depth, soil and moisture conditions improve steadily with each serai stage.

Primary plant succession is the development of vegetation on a 'natural' site (one that was not developed by humans). Where people construct a new site by destroying the current vegetation (as in shifting agriculture), plant recolonization will be more rapid and, in some ways, considerably different (since soil and moisture conditions will be more conducive from the start). This process is referred to as secondary plant succession (hence 'secondary forest').

Venn Diagram

A straightforward method of graphically expressing sets and subsets by drawing circles within rectangles. For example, if the rectangle represents a collection or network of central places, a circle within it may indicate the subset of central places that have at least one bank. Within a building society office, an additional subgroup might be identified, consisting of central locations. Because certain key locations will include both businesses, the two subset circles will be represented as crossing or overlapping on the Venn diagram.

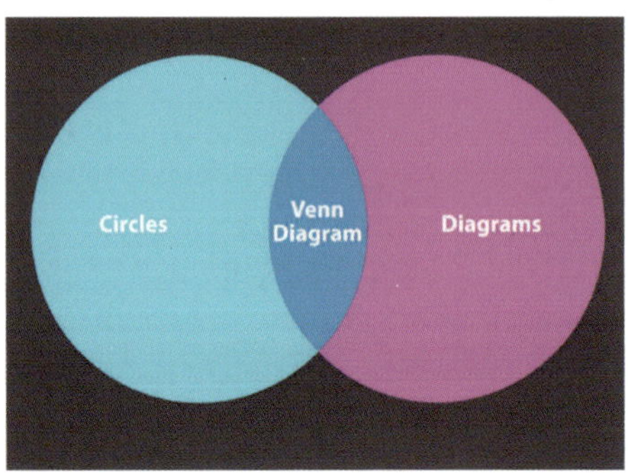

Figure 76. A basic Venn diagram.

Source: Image by flickr.com.

Ventifact

Reshaping and polishing of a stone by wind erosion in the desert. Initially, just the upwind side of the ventifact would be worn away, resulting in a facet or bevel. Other levels may occur if the wind direction changes periodically or the stone is flipped over. This is said to be the explanation for dreikanter (three-faceted stones).

Vertical Erosion

A river's in-situ down cutting in reaction to fast land uplift, sea-level fall, or an excess of energy owing to greater discharge without an equal addition of sediment load. Deep V-shaped valleys and incised meanders are formed as a result of vertical erosion. Lateral erosion of ct.

Figure 77. Vertical erosion of a river.

Source: Image by Wikimedia.

Vertical Expansion

One of three methods for a company to grow, in this case by engaging itself more in the same manufacturing cycle. For example, a brewing business may expand its sphere of activity by acquiring barley malting (*backward vertical growth*) or by controlling the distribution and retailing of its beer (*forward vertical expansion*).

Vertical Integration

Vertical integration occurs when multiple phases of the manufacturing process are located on the same site, as seen in integrated steelworks (where iron ore refinement and smelting may occur alongside sheet steel rolling) or a pulp and paper mill. Vertical integration allows possible saving time and transport expenses, which is a major advantage of having a series of manufacturing stages in one place, it also boosts sale economies.

Vicious Circle

This phrase is used in the context of regional development theory to explain the sequence of repercussions for the periphery of growing spatial concentration of resources and growth in the center. The term "*downward spiral*" is equally appropriate.

Village

A village is a collection of dwellings, which is smaller than a town but bigger than a hamlet. A Village consists of houses, farms, shops, places of worship, etc. However, it functions without a municipal government. Villages were often established as agricultural communities (though there are examples of planned *industrial villages*, such as New Lanark, Scotland, and Saltaire, Yorkshire), but this may no longer be the case, particularly for those located within the commuting circles of towns and cities (i.e., *Dormitory villages*).

Virtuous Circle

Myrdal (1957) coined the phrase to describe the cyclical process connected with the geographical concentration of resources in core locations, which allows them to preserve their original advantage. Sometimes the term "upward spiral" is used. This is known as a vicious circle.

Visible Trade

Importing and exporting products.

Vital Statistics

Births, marriages, deaths, and other documented facts regarding local, national, and worldwide populations are represented by numerical data.

Viticulture

Viticulture is a wide phrase that refers to the growth, preservation, and harvesting of grapes in an outdoor setting. It's usually for the production of wine.

Figure 78. Grapes in growing phase.

Source: Image by Pixabay.com.

Volcano

A conical hill or mountain (volcanic peak) built up by the ejection of igneous materials from a vent. The form of a volcano is influenced by many factors, such as the nature of the materials extruded (lava, cinders, or ash); the lava's chemical makeup (acid lava, which is rich in silica and very sticky, creates steep-sided, dome-like volcanoes; basic lava, which contains less quartz, is more mobile and forms more gently sloping volcanic cones); the extent to which either intense explosive activity or central collapse has produced a huge crater; and the volcano's following erosional history, which includes extensive dissection of its flanks by radial streams or, in certain cases, alteration by glacial erosion.

Some volcanoes originate along a fissure or fracture in the earth's crust. These are known as fissure volcanoes because they lack the unique form of individual volcanoes. Volcanoes can also be active, inactive, or extinct. Volumetric symbols are 'three-dimensional' symbols with volumes proportionate to the amount being visually represented (or cartographically, when located on a map). The cube root of the value being represented is used to build proportionate spheres, cubes, and columns.

Von Thunen's Model

A model of agricultural output and related land uses surrounding a market town was published in 1826 by a German economist-landowner. Several key simplifying assumptions are used in the model, including the fact that the market town is located in a physically homogeneous region and that transportation costs are directly proportional to distance is also believed that each farmer in the region would only sell his surplus produce in that town, that he will cover the whole expense of transportation alone, and that he will always seek to practice the sort of farming that will generate the most profit.

The model is based on the principle of economic rent, whereby different types of agricultural land use produce different net returns per unit area, because of different yields and different transport costs as result, the model is composed of several circular zones, each of which is distinguished by a certain type of agricultural output. Because they demand the greatest work, have the highest transportation costs, and produce a perishable commodity, market gardening and dairying are placed closest to the market town.

They produce the largest return per unit area, but as the distance from the market grows, the return decreases to the point where it is more profitable to pursue another form of agriculture. Agriculture's profitability and intensity continue to diminish outward from the market, resulting in one form of farming succeeding another until cattle-grazing is achieved.

Stock grazing becomes the major industry at the region's perimeter since it needs the least labor and has the lowest transportation expenses (the cattle is brought to market on the hoof). Despite the fact that technology and market conditions have changed dramatically since the early nineteenth century, and that some of von Thunen's simplifying assumptions must be questioned, the model remains useful, if only that it links economic concepts with spatial locations.

V-Shaped Valley

In contrast to a glaciated U-shaped valley, a 'normal' river valley. The specific angle of the V will be determined by the respective rates of vertical river erosion and valley slope recession caused by weathering, rain-wash, and mass movement. If river down cutting is relatively quick, the V will be narrow, but if slope recession is quite rapid (as in a non-resistant rock area or a humid location), the V will be wide. The V will be wider. Not all river valleys have a V-shaped cross-section. The valley will be flat-floored when lateral stream erosion is effective. A comparable impact will occur if there is a substantial deposition of alluvium on the valley floor, such as, as a result of an increase in the base level of erosion.

Vulcanicity

The mechanisms through which solid, liquid, and gaseous constituents are either pushed into or escapes from the earth's crust via vents and fissures. As a result, the phrase encompasses igneous activity in general, rather than only the creation of volcanoes.

W

Wadi

In a hot desert, a steep-sided, flat-floored valley with a dry floor. Wadis have many of the characteristics of typical river valleys (dendritic patterns and gently concave long-profiles) and are periodically inhabited by running water under current conditions. There is no doubt, however, that the wadis of the Sahara and Arabian deserts were mostly eroded in the recent past when deserts experienced extreme weather.

Warm Front

The well-defined border between a warm and cold air mass, when the former is advancing and overriding the latter, such as at the leading edge of a depression's warm sector. The gradient of the warm front is significantly less steep than that of the cold front and the comparatively sluggish rise of warm wet air at the front results in cooling and condensation across a broad area, where the formation of nimbus clouds happens. A warm front's passage is generally defined by a protracted spell of steady rain that lasts many hours.

Warm (Warm-Based) Glacier

Temperatures near 0°C are seen throughout the depths of a glacier (pressure melting point). Warm glaciers are also known as temperate glaciers since they are found in mid-latitude mountain ranges such as the Alps and Rockies. Warm glaciers create a considerable amount of melt water in the summer, which enters the glaciers through moulins and crevasses, producing an englacial and subglacial drainage system (which is absent from cold glaciers). Melt water at the foot of a warm glacier causes the ice to slide over bedrock, promoting efficient abrasion. In winter, the upper few meters of a warm glacier may be chilled below 0°C, but the major part of the ice remains at, or close to, pressure melting point.

Figure 79. Warm glacier in the western pacific warm pool.

Source: Image by Imaggeo.

Warm Sector

A 'wedge' of warm, moist air confined within a mid-latitude frontal depression, tapering northwards in the northern hemisphere and southwards in the southern hemisphere. The warm front (with its constant cloud cover and long duration of rainfall) is ahead of the warm sector, while the cold front is behind it (with its cumulus clouds and heavy showers). The warm sector causes an interval of pleasant, moderate weather, with 'fair weather' cumulus clouds and bright spells. The area of the warm sector shrinks with time as a result of the cold front's considerably faster forward movement, which gradually overtakes the warm front.

Waste Disposal

It is a process of getting rid of trash or residues (liquid or solid) generated by human consumption of resources. In recent times, garbage is disposed of at the following locations Landfills, mainly for household trash and certain chemical waste; rivers for treated sewage water and low-level, non-toxic chemical waste. Garbage; is becoming all sorts of excess waste – a rising worldwide problem; import waste from medics and are paid enormous sums of money for doing so, in violation of international law. The safe disposal of nuclear waste is a contentious subject right now.

Water Balance

How evapotranspiration, run-off, and changes in the amounts of water stored in soil and rocks account for precipitation received at a location (groundwater). The water balance (also known as the hydrological balance budget) is computed using the formula: P=E+R+S.

Where P denotes precipitation, E denotes evapotranspiration, R denotes run-off, and S denotes changes in storage throughout the course of the measurement (usually one year). If a study of water balance is conducted over time, it may become, for all intents and purposes, constant and is therefore sometimes removed from the water balance equation. It is critical to understand that large changes in water balance can occur within the area. It is critical to recognize that major changes in water balance can occur within a year. In the United Kingdom, for example, a significantly higher fraction of precipitation is lost to run-off in the winter than in the summer, whereas evapotranspiration is greatest in the summer (when it may surpass precipitation) but inconsequential in the winter. Furthermore, water tends to flow into storage during the winter (when subterranean water is replenished by percolation), but out of storage during the summer to supply river base flow.

Waterfall

When a river course is disrupted by a marked break, such as the edge of a plateau or mountain, from where waterfalls vertically or in a step-like manner depending on the terrain, is categorized as a waterfall. It is at the junction of a valley with a glacial trough or can be found at sea cliffs, rarely. Some waterfalls are formed as a result of differential erosion; for example, along a fault line, soft rocks may be pushed up against strong rocks and rapidly eroded by fluvial activity, resulting in the development of a cascade at the fault. Even if waterfalls are begun by structures such as faults, they can be separated from them by the process of head ward erosion.

Water Gap

A valley through a cuesta or ridge that has been eroded by a river, that has continued to occupy the gap.

Water Table

The zone of saturation's top surface is porous rock. When precipitation exceeds evapotranspiration, rainwater percolates to the water table (which can be found at depths of tens or even hundreds of meters under the surface). In the United Kingdom, this happens mostly in the winter, causing the water table to rise to a maximum height in the spring.

Percolation effectively ends in the summer, and the saturation zone is drained by springs and seepages (which develop where the water table intersects the land surface, at the base of scarps, and in deep valley bottoms). The slope of the water table varies from place to place (it is steeper where the rock is less permeable and softer where the rock is extremely porous) and, in general, follows the form of the surface relief in a muted way, but there are notable exceptions to this norm. Where the percolation of rainfall to the main water table is locally obstructed (e.g., by a small clay layer), secondary regions of saturation, or perched water tables, occur.

Watershed

The line that separates the headstreams that flow into various drainage basins; effectively, the line that defines the margin of a drainage basin.

Watersheds can be defined abruptly (by the crest of a hill) or indeterminately (in low-lying places where the 'split' between river basins is broad and soft).

Figure 80. Watershed illustration.

Source: Image by flickr.com.

Wave

A surface feature of oceans, seas, and inland waters that consists of a linear crest separating parallel troughs. Winds sweeping across the sea cause the majority of the waves to form. The flowing air creates a frictional pull on the surface water particles, causing a sequence of orbital water motions. These are 'ahead' during the wave crest and 'backward' in the trough, causing oscillations. The entire waveform travels in the direction of airflow. Waves can be precisely described in terms of height – the vertical distance between crest and trough; length – the distance between two successive crests; period – the time required for a wave to advance by one wavelength; and frequency.

There is a basic contrast between long, low waves and short, steep waves. Wind speed, wind duration, and distance from a lee coast are the primary influences on waveform and size. When waves reach the beach, they break when the orbital velocity of the water particles exceeds the wave's forward velocity, which is lowered by friction caused by the sea bed in shallow water.

Wave-Cut Platform

At the base of a retreating sea cliff, a near-level surface worn-in solid rock by wave action is a wave cut platform. The breadth of wave-cut platforms varies from a few meters to hundreds of meters, depending on the rock type and resistance, as well as the length of marine erosion at its current level. Wave-cut platforms vary greatly in detail. Some are very irregular, with eroded grooves and depressions along faults, joint lines, and weak layers. Wave-cut platforms along Britain's shoreline frequently appear to be 'composite,' with two or three different levels which are a few meters apart vertically, some small variations are represented in sea level over the late Pleistocene and post-glacial eras.

Wave Refraction

When waves change their route, as they are reaching headlands or beaches, as well as they cross distal ends of spits and bars, this process is known as wave refraction. In these cases, refraction is caused by the shallowing of the bottom, which has the effect of lowering wave velocity. Thus, waves that approach the beach obliquely are 'turned,' such that when wave-break occurs, their crests are almost parallel to the land. The impact of refraction around headlands is to concentrate wave energy (shown by orthogonal lines) and therefore erosive potential on the promontory, which becomes cliffed, creating arches, stacks, and stumps, and to limit wave energy in the intervening bays, where deposition prevails to build beaches. A refraction may be primarily responsible for the formation of re-curved tips at the far ends of spits.

Weathering

The breakdown and disintegration of rocks in situ, resulting in a waste mantle or loose debris that can be transported by transport mechanisms. There are two forms of weathering: chemical weathering and mechanical (physical) weathering. However, biological or organic weathering is sometimes considered as a third kind.

Figure 81. Erosion of rocks weathering.

Source: Image by Pixabay.com.

Weber's Theory of Industrial Location

This hypothesis, published in 1909, contends that companies gravitate toward low-cost locations and that such places are usually points of lowest transportation costs. As with most theories, Weber made many simplifying assumptions, such as terrain uniformity, the fact that some raw materials occur in fixed locations while others occur everywhere, standardized wage rates, the existence of perfect competition and economic man, and that transportation costs are determined by the weight of the load and distance.

The picture below depicts the basic scenario of two raw-material sources (rt and r2) and a single market (m). If transportation costs are assumed to be the same for both raw materials and finished goods, the theory predicts that the least-cost site will be at the center of the locational triangle (p,), equidistant from r1, r2, and m. Because those materials effectively contain waste, the lowest-cost site will be closer to the raw material sources (at p2 rather than p1). In contrast, if an industry is gaining weight, the lowest-cost location will be closer to the market (at p3 rather than p1).

Weber's thesis has been attacked on many fronts, most of which are related to his basic assumptions. Weber eventually admitted to the necessity to account for things like geographical differences in labor availability and labor prices, as well as the magnetic influence of existing agglomerations on an industrial placement. Nonetheless, the theory's major critique is that it places an excessive focus on transportation costs and wrongly believes that such costs are directly proportional to distance and weight.

Weight-Loss Ratio

It is the difference, between the weight of raw materials utilized in a manufacturing process and the weight of the finished product. In an industrial site, the weight-loss ratio is thought to be important since the greater the ratio, the more probable production will be situated near raw material sources.

Welfare

Firstly, Welfare can be described as the overall status or condition of society, as in a welfare state. It is a relatively recent focus point in human geography and encompasses factors like nutrition, housing, medical care, education, work, and so on, all of which contribute to human pleasure.

There is a distinction to be made between economic and social wellbeing former generally refers to what individuals get from the consumption of goods and services, whereas the latter covers all of the elements that contribute to the quality of human life. There appears to be some overlap or confusion with quality of life in this sense, but one possible clarification would be to view welfare as only one essential facet of it. Secondly, welfare in the United States means additional benefits provided to the households struggling to survive, by the government.

Welfare Geography

A subfield of human geography emerged in the 1970s when interest in the subject changed from model-building and quantification to a rising concern with broad problems such as quality of life, territorial justice, and welfare. It focuses on current issues such as poverty, starvation, crime, and disparities in access to housing and social services. One of the most prominent proponents of welfare geography defines it as "who receives what, where, and how." There is little question that marxism has inspired much welfare geography, and that it has a strong radical element.

Well-Being

A catch-all word for a collection of overlapping ideas such as standard of living, level of living, quality of life, social satisfaction, and welfare.

Wetland

Those lands that are occasionally or continuously flooded, such as tidal mudflats in an estuary, a marsh, or a bog. Wetlands are essential feeding and nesting places for birds, and many are under threat of reclamation, as seen in Hong Kong, Japan, and Singapore.

Figure 82. Scientists studying wetlands.

Source: Image by Phys.org.

Wholesaling

It is a service that is processed before a product reaches the retailers or customers. It takes place between the manufacturer and the retailer/ shop owner. The major tasks of wholesaling include breaking of bulk; warehousing (keeping inventories to meet variations in demand); assisting in financing distribution by granting credit to retailers; and, in certain cases, preparing a product for sale by grading, packaging, and branding.

Wilderness

In conservation, it refers to a region that has been kept undisturbed in its natural condition, with little or no human management and intervention (as, for example, mountains and large areas of desert, of taiga and tundra). The recreational potential of such locations is rapidly being appreciated, since they provide a variety of options ranging from adventure to birding, camping to exploration, or simply the chance to relax.

Wind Chill

The influence of wind on temperature,' causes it to seem extremely cold even when the average temperature is well above freezing. Wind chill affects people's susceptibility to exposure because it accelerates the rate of loss of body heat, hastening the onset of hypothermia.

Windbreak

To offer protection for houses, soil, crops, or animals, a barrier, generally a hedge or belt of trees, is placed across the direction of the prevailing wind. Windbreaks are useful in exposed locations, such as the Netherlands, France's Rhone valley, and portions of Australia, to minimize soil erosion and wind damage to crops.

Wind Power

One of the few commercially viable renewable energy sources being developed on a significant scale. Gathering and fitting wind turbines in the wind farms with exposure to simple amounts of wind could develop as wind farms, generating wind power.

Figure 83. Wind turbines generate wind power.

Source: Image by pxhere.com.

Word Bank

The World Bank, formerly known as the International Bank for Reconstruction and Development, is a United Nations organization. It was established in 1945 to assist in post-war reconstruction throughout Europe Nonetheless, it then shifted its focus to aiding the underdeveloped globe. In the latter case, it has given loans totaling US$203 billion to third-world countries. Despite their good intentions, these loans have contributed to the growth of third-world debt. This vulnerability has been mitigated to some extent by the shift toward interest-free financing.

WWF (Worldwide Fund for Nature)

Is an NGO formed in 1961, to create and spread awareness about the protection of endangered species? Also aims at addressing significant dangers to the environment.

X

Xenophobia

Fear, dislike, or distrust of foreigners or strange things caused by the fear in the individual

Xerophyte

A plant that has evolved to tolerate seasonal or persistent drought. Xerophytes (such as desert cacti and thorn plants on desert borders) have unusually long roots (to access groundwater), thick bark, tiny glossy leaves (to decrease evaporation), and the ability to store water when it becomes available (like succulents, hold moisture in a spongy substance in their stems).

Xerosere

A plant succession that evolves in a dry environment, such as a bare rock surface or a dune.

Y

Yardang

Wind erosion creates a desert landform. Yardangs are elongated ridges created parallel to the prevailing wind direction, with evident evidence of basal undercutting caused by the contact of wind-borne sand grains, resulting in a 'blasting effect' on their upwind sides alone.

Figure 84. Yardang: a streamlined protuberance carved from bedrock.

Source: Image by flickr.com.

Yield

First, the rate of return on capital investment over a given period, generally stated in percentage terms. Second, output or production measured in proportion to one of the inputs, such as cereal production per hectare or factory production per man-hour.

Young Fold Mountains

Mountain shaped like curves formed by earth movements during the mid-tertiary alpine orogeny. Young fold mountains have high heights (as seen in the Himalayas, Andes, and Alps), partially due to their young age and lack of time for planation to occur, and in part to on-going isotactic uplift caused by the folding motions. Their relief is very uneven as a result of both deep vertical incision by rivers and extensive glacial erosion, primarily during the Pleistocene epoch – but glaciation is currently occurring in many young fold mountains.

Yuppy

It is technically not considered as an acronym, but it stands for 'young, upwardly, mobile, person, i.e., Someone who has successfully launched a dynamic profession and enjoys a high level of financial pay; most likely someone who is in stage one of the life cycle.

Z

Zero Population Growth

Zero population growth, also known as the replacement level of fertility, is a situation of demographic balancing in which the number of individuals in a given population neither rises nor decreases. It is a state that many doctors are beginning to consider as a result of a significant decrease in fertility. Without a doubt, it will need the addition of a new stage to the demographic transition.

Zeugen

Tabular masses of hard sedimentary rock (typically sandstone) formed in deserts by selective wind erosion. Zeugen can reach a height of 30 meters and is separated from one another by depressions 'scoured out' by the wind, which has been able to attack and destroy weak underlying shales.

Zonal Soil

Type of a soil that undergoes extensive development. It has been influenced by soil-forming processes such as humification, leaching and eluviation, and acidity over a long period of time, resulting in a well-developed soil profile and a substantially decreased impact on the parent material. As a result, zonal soils reflect wide climatic and vegetative influences (as in the case of pod sols developed in cool temperate climates where precipitation is adequate for coniferous forest growth). Soil categorization on a global scale is based on the identification of zonal soils. Also, refer to azonal and intrazonal soils.

Figure 85. Zonal soil during December season.

Source: Image by flickr.

Zone of Assimilation

A transitory zone formed by the shifting front of a CBD, typically involving the invasion of residential neighborhoods by various forms of core business (notably small offices and retailing firms). Two subzones may be identified in the transition from the real front of the zone to the 'core of the CBD':

i. In an early phase in which residences are simply converted into commercial premises; and
ii. A redevelopment and consolidation phase closer to the center, in which converted dwellings are gradually replaced by structures purpose-built to suit central enterprises. This second stage usually occurs when real estate investors and property developers are persuaded that the danger connected with CBD migration has been minimized to an acceptable level.

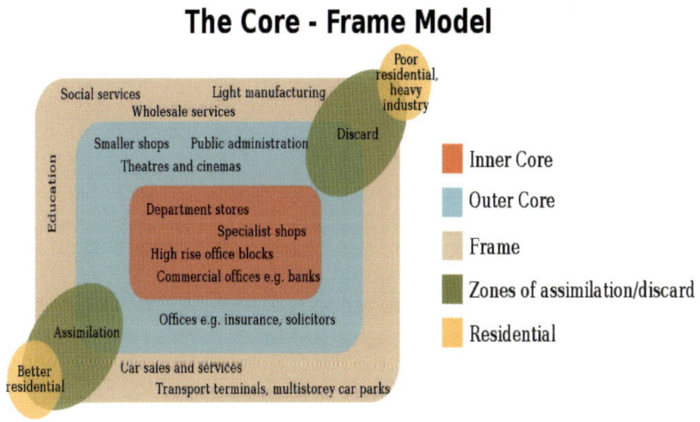

Figure 86. Core frame model.

Source: Image by Wikimedia.

Zone of Discard

When a core business attempts to relocate, a transitory zone is formed following the shifting process of CBD. When buildings are left abandoned. Keep your location close to the CBD's changing center of gravity. The degree of withdrawal rises in the opposite direction as the CBD is traveling. When central commercial enterprises' desertion is complete, chances for conversion of existing premises to new users or total redevelopment of the region to make room for new activity appear.

Zone of Transition

The second ring of the city structure's concentric zone scheme. Although it was initially a residential district, its proximity to the city Center makes it appealing for commercial and industrial growth, especially as the CBD develops This happening that takes place due to the invasion of non-residential activities, along with the natural aging of houses and urban infrastructure, eventually leads to a fall in residential attractiveness.

As households migrate in pursuit of better housing and more appealing residential settings elsewhere in the developing metropolis, poorer households, and ethnic minorities take their place, causing the city's population to rise, forming a ghetto. The areas which are subdivided, causing overcrowding and further deterioration of the places are known as dwellings. As the structure and atmosphere deteriorate, the zone becomes more characterized by high levels of crime. Consider the twilight.

BIBLIOGRAPHY

1. A Level Geography. N.d. Carbon Cycle. [online] Available at: https://www.alevelgeography.com/carbon-cycle/ (accessed 9 September 2021).
2. Ackerman, E., (1953). Regional Research--Emerging Concepts and Techniques in the Field of Geography. *Economic Geography*, [online] *29*(3), 189. Available at: https://www.jstor.org/stable/142516 (accessed 9 September 2021).
3. Alexander, R., (2007). Geography and soil properties by A. F. Pitty. Methuen, London, (1979). No. of pages: 287. Price: £5.50 (paperback). *Geological Journal*, [online] *19*(4), 402–402. Available at: https://onlinelibrary.wiley.com/doi/abs/10.1002/gj.3350190409 (accessed 9 September 2021).
4. Anthamatten, (2010). Spatial Thinking Concepts in Early Grade-Level Geography Standards. *Journal of Geography*, [online] *109*(5), 169–180. Available at: https://www.researchgate.net/publication/233307361_Spatial_Thinking_Concepts_in_Early_Grade-Level_Geography_Standards (accessed 9 September 2021).
5. Bacon, R., & Green, J., (1981). Core Concepts in Introductory Physical Geography. *Journal of Geography*, [online] *80*(3), 104–108. Available at: https://people.uwec.edu/ivogeler/w401-capstone/coreconceptsphysicalgeography.pdf (accessed 9 September 2021).
6. Bacon, R., (1979). Building a curriculum in introductory human geography through core concepts. *Journal of Geography*, [online] *78*(4), 152–156. Available at: https://in.booksc.eu/book/31585558/1fcc24 (accessed 9 September 2021).

7. Bathelt, H., & Glückler, J., (2005). Resources in Economic Geography: From Substantive Concepts towards a Relational Perspective. *Environment and Planning A: Economy and Space*, [online] *37*(9), 1545–1563. Available at: https://www.researchgate.net/publication/23539585_Resources_in_Economic_Geography_From_Substantive_Concepts_towards_a_Relational_Perspective (accessed 9 September 2021).
8. Baudet-Michel, S., (2012). Key Concepts in Economic Geography. By Yuko Aouama, James T. Murphy & Susan Hanson, London 2011: Sage. 278 pp. Paperback, ISBN: 9781847878953, £21.99. Hardcover, ISBN: 9781847878946, £67.00. *Tijdschrift voor economische en sociale geografie*, [online] *103*(5), 626–629. Available at: https://zh.mindat.org/reference.php?Id=571354 (accessed 9 September 2021).
9. Bennett, K., & Prowse, T., (2009). Northern Hemisphere geography of ice-covered rivers. *Hydrological Processes*, [online] p.n/a-n/a. Available at: https://onlinelibrary.wiley.com/doi/abs/10.1002/hyp.7561 (accessed 9 September 2021).
10. Blyn, G., (1960). A diagram for demonstrating certain concepts in economic crop geography. *The Professional Geographer*, [online] *12*(2), 1–3. Available at: https://onlinelibrary.wiley.com/doi/abs/10.1111/j.0033–0124.1960.122_1.x (accessed 9 September 2021).
11. Boyle, M., (2001). Towards a (re)theorization of the historical geography of nationalism in diasporas: the Irish diaspora as an exemplar. *International Journal of Population Geography*, [online] *7*(6), 429–446. Available at: https://onlinelibrary.wiley.com/doi/abs/10.1002/ijpg.240 (accessed 9 September 2021).
12. Brenchley, K., (2001). World regional geography by Lydia Mihelic Pulsipher, W. H. Freeman and company, London, (1999). No. of Pages: 546. Price: £25.95. ISBN 0–7167–3092–8. *Geological Journal*, *36*(1), 89–90.
13. Butler, T., (2011). Key concepts in urban geography – By Alan Latham, Derek McCormack, Donald McNeill and Kim Macnamara. *Area*, [online] *43*(2), 239–239. Available at: https://www.researchgate.net/publication/229826906_Key_concepts_in_urban_geography_-_By_Alan_Latham_Derek_mccormack_Donald_mcneill_and_Kim_macnamara (accessed 10 September 2021).
14. Cantwell, J., & Mudambi, R., (2011). Physical attraction and the geography of knowledge sourcing in multinational enterprises. *Global*

Strategy Journal, [online] *1*(3–4), 206–232. Available at: https://onlinelibrary.wiley.com/doi/abs/10.1002/gsj.24 (accessed 9 September 2021).

15. Carrying Capacity (2021). [online] Available at: https://www.biologyonline.com/dictionary/carrying-capacity (accessed 9 September 2021).
16. Chester, D., (1987). Statistics in geography by David Ebdon, Basil Blackwell, (1985). No. of pages: 232. Price: £8.50 (soft covers). *Geological Journal*, [online] *22*(1), 61–62. Available at: https://onlinelibrary.wiley.com/doi/abs/10.1002/gj.3350220108 (accessed 9 September 2021).
17. Chester, D., (2007). Horizons in Physical Geography edited by M. J. Clark, K. J. Gregory and A. M. Gurnell, Macmillan Education, (1987). No of pages: 395. Price: £30 (hardcover). *Geological Journal*, [online] *24*(3), 234–234. Available at: https://onlinelibrary.wiley.com/doi/full/10.1002/gj.3350240313 (accessed 9 September 2021).
18. Chouinard, V., (1997). Structure and agency: contested concepts in human geography. *The Canadian Geographer/Le Géographe canadien*, [online] *41*(4), 363–377. Available at: https://www.researchgate.net/publication/229590411_Structure_and_agency_Contested_concepts_in_human_geography (accessed 10 September 2021).
19. Cox, K., (1995). Concepts of space, understanding in human geography, and spatial analysis. *Urban Geography*, [online] *16*(4), 304–326. Available at: https://www.tandfonline.com/doi/abs/10.2747/0272–3638.16.4.304 (accessed 9 September 2021).
20. Cultural geographies, (2015). Book Review: Key Concepts in Historical Geography. [online] *22*(2), 380–380. Available at: https://journals.sagepub.com/doi/abs/10.1177/1474474014539252 (accessed 9 September 2021).
21. Cutchin, M., (1999). Qualitative Explorations in Health Geography: Using Pragmatism and Related Concepts as Guides. *The Professional Geographer*, [online] *51*(2), 265–274. Available at: https://r.search.yahoo.com/_ylt=Awr9DuHv.9vggycaa0nxnyoa;_ylu=y29sbwnncteecg9zazeednrpzamec2vja3ny/RV=2/RE=1624665199/RO=10/RU=https%3a%2f%2fwww.tandfonline.com%2fdoi%2fabs%2f10.1111%2f0033–0124.00163/RK=2/RS=hanxcwkfilgsv0bgy2df_Y_cfhs- (accessed 10 September 2021).

22. De Ploey, J., (1980). Soil erosion, R. P. C. Morgan, Topics in Applied Geography series, Longman. No. of pages: 113. Price: £3.95. *Earth Surface Processes*, [online] *5*(2), 201–202. Available at: https://onlinelibrary.wiley.com/doi/abs/10.1002/esp.3760050215 (accessed 9 September 2021).
23. Drinkwater, L., Friedman, D., & Buck, L., (2021). Understanding Agricultural Systems – SARE. [online] SARE. Available at: https://www.sare.org/publications/systems-research-for-agriculture/chapter-one-introduction-to-agricultural-systems-and-agricultural-systems-research-a-paradigm-change/understanding-agricultural-systems/ (accessed 9 September 2021).
24. Edsall, R., & Wentz, E., (2007). Comparing Strategies for Presenting Concepts in Introductory Undergraduate Geography: Physical Models vs. Computer Visualization. *Journal of Geography in Higher Education*, [online] *31*(3), 427–444. Available at: https://www.tandfonline.com/doi/abs/10.1080/03098260701513993 (accessed 9 September 2021).
25. Ehlers, J., (1993). Quaternary environment in Hungary edited by M. Pécsi and F. Schweitzer, Studies in Geography in Hungary, 26, Akadémai Kiadó, Budapest, (1991). No. of pages: 103. Price: $15.00. ISBN 963–05–6080–1. *Earth Surface Processes and Landforms*, [online] *18*(8), 758–759. Available at: https://onlinelibrary.wiley.com/doi/abs/10.1002/esp.3290180813 (accessed 9 September 2021).
26. Encyclopedia Britannica. (2017). Cartography | Geography. [online] Available at: https://www.britannica.com/science/cartography (accessed 9 September 2021).
27. Encyclopedia Britannica. (2021). Agglomerate | Rock. [online] Available at: https://www.britannica.com/science/agglomerate (accessed 9 September 2021).
28. Encyclopedia Britannica. (2021). Agricultural revolution | English history. [online] Available at: https://www.britannica.com/topic/agricultural-revolution (accessed 9 September 2021).
29. Encyclopedia Britannica. (2021). Calcrete | Geology. [online] Available at: https://www.britannica.com/science/calcrete (accessed 9 September 2021).
30. Encyclopedia Britannica. (2021). Carnivore | Consumer. [online] Available at: https://www.britannica.com/science/carnivore-consumer (accessed 9 September 2021).

31. English, (1968). Landscape, Ecosystem, and Environmental Perception: Concepts in Cultural Geography. *Journal of Geography*, [online] *67*(4), 198–205. Available at: https://www.researchgate.net/publication/352648117_From_preference_to_landscape_sustainability_A_review_on_landscape_preference_research_from_1968_to_2019 (accessed 9 September 2021).
32. Exploring Spatial Scale in Geography, (2014). Scale in Spatial Data Analysis: Key Concepts. [online] pp.9–28. Available at: https://onlinelibrary.wiley.com/doi/10.1002/9781118526729.ch2 (accessed 9 September 2021).
33. Fabbri, D., & Robone, S., (2010). The geography of hospital admission in a national health service with patient choice. *Health Economics*, [online] *19*(9), 1029–1047. Available at: https://pubmed.ncbi.nlm.nih.gov/20648520/ (accessed 9 September 2021).
34. Fouberg, E., (2013). "The world is no longer flat to me": student perceptions of threshold concepts in world regional geography. *Journal of Geography in Higher Education*, [online] *37*(1), 65–75. Available at: https://adamdastrup.weebly.com/uploads/1/7/4/1/17412073/threshold_concepts_in_world_geography.pdf (accessed 9 September 2021).
35. Fouberg, E., (2016). Reflecting on threshold concepts in world regional geography. *Journal of Geography in Higher Education*, [online] *40*(1), 9–13. Available at: https://www.tandfonline.com/doi/full/10.1080/03098265.2016.1131903 (accessed 9 September 2021).
36. Geoarchaeology, (1989). 26th Congress of the International Geographical Union, held in Sydney, Australia, 22–26 August (1988). D. Dragovich, Department of Geography, University of Sydney, Sydney 2006, Australia. [online] *4*(3), 271–272. Available at: https://onlinelibrary.wiley.com/doi/abs/10.1002/gea.3340040305 (accessed 9 September 2021).
37. Geoarchaeology, (1995). Geological Society of America, Annual Meeting, Boston, Massachusetts. William C. Johnson, Department of Geography, University of Kansas, Lawrence, Kansas 66045. (October 25–28, 1993). [online] *10*(2), 151–154. Available at: https://onlinelibrary.wiley.com/doi/abs/10.1002/gea.3340100209 (accessed 9 September 2021).
38. Geological Journal, (2007). Physical Geography of Asiatic Russia, by S. P. Suslov, translated from the Russian by N. D. Gershevsky. 600

pp., numerous photographs and maps. W. H. Freeman and Co. Ltd., London, (1961). Price 105/-. [online] *3*(1), V-V. Available at: https://onlinelibrary.wiley.com/doi/abs/10.1002/gj.3350030115 (accessed 9 September 2021).

39. Giardino, J., & Fitzgerald, J., (1987). The Encyclopaedic Dictionary of Physical Geography, Andrew Goudie, B. W. Atkinson, K. J. Gregory, I. G. Simmons, D. R. Stoddart, and David Sugden, Eds., 1985, Oxford and New York: Basil Blackwell, 528 pp., Maps, Diags., Photos, Refs., & Index, $60.00 (ISBN 0–631–13292–9). *Geoarchaeology*, [online] *2*(2), 163–163. Available at: https://onlinelibrary.wiley.com/doi/abs/10.1002/gea.3340020206 (accessed 9 September 2021).
40. Gorkin, A., (2011). On the relativity of indicators and concepts in socioeconomic geography. *Regional Research of Russia*, [online] *1*(2), 157–164. Available at: https://link.springer.com/article/10.1134/S2079970511020031 (accessed 10 September 2021).
41. Graham, E., & Boyle, (2001). Editorial introduction: (re)theorising population geography: mapping the unfamiliar. *International Journal of Population Geography*, [online] *7*(6), 389–394. Available at: https://onlinelibrary.wiley.com/doi/abs/10.1002/ijpg.237 (accessed 9 September 2021).
42. Hahvey, D., (1966). Theoretical concepts and the analysis of agricultural land-use patterns in geography. *Annals of the Association of American Geographers*, [online] *56*(2), 361–374. Available at: https://onlinelibrary.wiley.com/doi/full/10.1111/j.1467–8306.1966.tb00563.x (accessed 10 September 2021).
43. Hamilton, A., & Martin, R., (2010). Inferring semantic organization from refractory access dysphasia: Further replication in the domains of geography and proper nouns but not concrete and abstract concepts. *Cognitive Neuropsychology*, [online] *27*(8), 614–635. Available at: http://doi.org/10.1080/02643294.2011.609541 (accessed 9 September 2021).
44. Hansen, H., (2012). Key Concepts in Economic Geography. *European Planning Studies*, [online] *20*(5), 893–894. Available at: https://www.tandfonline.com/doi/pdf/10.1080/09654313.2012.667931 (accessed 9 September 2021).
45. Hardwick, S., (2003). Migration, embedded networks and social capital: towards theorising North American ethnic geography. *International*

Journal of Population Geography, [online] 9(2), 163–179. Available at: https://onlinelibrary.wiley.com/doi/abs/10.1002/ijpg.277 (accessed 9 September 2021).

46. Hartshorne, R. D., (1939). Generic concepts and principles in geography. *Annals of the Association of American Geographers*, [online] *29*(4), 378–397. Available at: https://www.tandfonline.com/doi/abs/10.1080/00045603909357329 (accessed 9 September 2021).
47. Investopedia. (2020). Capitalism. [online] Available at: https://www.investopedia.com/terms/c/capitalism.asp (accessed 9 September 2021).
48. Jones, A., (2012). Key Concepts in Economic Geography – By Yuko Aoyama, James Murphy, and Susan Hanson. *Economic Geography*, [online] *88*(2), 225–226. Available at: https://onlinelibrary.wiley.com/doi/abs/10.1111/j.1944–8287.2011.01144.x (accessed 10 September 2021).
49. *Journal of Hydrology*, (1971). Small watershed experiments; an appraisal of concepts and research developments. 14(3–4), 360.
50. Kawai, M., (1987). Backwardation. The New Palgrave Dictionary of Economics, [online] pp.1–2. Available at: https://link.springer.com/referenceworkentry/10.1057/978–1–349–95121–5_444–1 (accessed 9 September 2021).
51. Kindon, S., (2013). Key concepts in development geography by Rob Potter, Dennis Conway, Ruth Evans and Sally Lloyd-Evans, Sage Publications Ltd., London. (2012). 278 pp. ISBN 978–0–85702–585–2. *New Zealand Geographer*, [online] *69*(2), 168–169. Available at: https://onlinelibrary.wiley.com/doi/abs/10.1111/nzg.12019_2 (accessed 9 September 2021).
52. Krishnamurthy, S., (1992). The geography of non-ocular malignant melanoma in India: its association with latitude, ozone levels and UV light exposure. *International Journal of Cancer*, [online] *51*(2), 169–172. Available at: https://pubmed.ncbi.nlm.nih.gov/1568786/ (accessed 9 September 2021).
53. Leszczycki, S., (1960). The Latest Approaches and Concepts in Geography. *Soviet Geography*, 1(4), 3–17.
54. Mandel, R., (1994). Environmental change: Contemporary problems in geography, 3rd edition. Andrew Goudie, 1992, Oxford University Press, xx + 329 pp., $55.00 (hardbound), $23.00 (paperbound). *Geoarchaeology*, [online] 9(6), 518–520. Available at: https://

onlinelibrary.wiley.com/doi/abs/10.1002/gea.3340090607 (accessed 9 September 2021).

55. Martinson, T., (1984). Using Guided Design to Teach Behavioral Concepts in Cultural Geography. *Journal of Geography*, [online] *83*(2), 79–82. Available at: https://onlinelibrary.wiley.com/doi/full/10.1111/j.1467–9663.1971.tb01147.x (accessed 9 September 2021).
56. Mawdsley, E., (2013). Key Concepts in Development Geography. By Potter, R., D. Conway, R. Evans & S. Lloyd-Evans, London 2012: Sage. 288 pp. ISBN 978–0–85702–585–5. *Tijdschrift voor economische en sociale geografie*, 104(2), 253–254.
57. Melachroinos, K., (2013). Key concepts in economic geography. Aoyama, Yuko, Murphy, James T., & Hanson, Susan. *Geografiska Annaler: Series B, Human Geography*, [online] *95*(2), 209–210. Available at: https://www.tandfonline.com/doi/abs/10.1111/geob.12019 (accessed 9 September 2021).
58. Merriam, D., (1979). Concepts and Techniques in Modern Geography (CATMOG): Study group in quantitative methods of the Institute of British Geographers. *Computers & Geosciences*, [online] *5*(1), 139. Available at: https://www.sciencedirect.com/journal/computers-and-geosciences/vol/5/issue/1 (accessed 10 September 2021).
59. Metoyer, S., Bednarz, S., & Bednarz, R., (2015). Spatial Thinking in Education: Concepts, Development, and Assessment. *Geospatial Technologies and Geography Education in a Changing World*, [online] pp.21–33. Available at: https://link.springer.com/chapter/10.1007/978–4–431–55519–3_3 (accessed 9 September 2021).
60. Mishra, (2021). Cash crops in India and their importance in The Indian economy – Utkal Today. [online] Utkal Today. Available at: https://www.utkaltoday.com/cash-crops-in-india/ (accessed 9 September 2021).
61. Missimer, T., Al-Mashharawi, S., Dehwah, A., & Coulibaly, K., (2017). Backreef and beach carbonate sediments of the Red Sea, Saudi Arabia: impacts of reef geometry and currents on sediment composition. Coral Reefs, [online] *36*(4), 1157–1169. Available at: https://link.springer.com/article/10.1007/s00338–017–1607–4 (accessed 9 September 2021).
62. Mostinsky, I., n.d. ADSORPTION. A-to-Z Guide to Thermodynamics, Heat and Mass Transfer, and Fluids Engineering, [online] Available at:

https://thermopedia.com/content/292/ (accessed 9 September 2021).

63. Paulissen, E., (1990). The physical geography, geomorphology and late quaternary history of the mahidasht project area, Qara Su Basin, Central West Iran, Ian A. Brookes, 1989, Royal Ontario Museum Publications in Archaeology, ROM Mahidasht Project Volume 1, 5 plates, xi + 48 pp., $21.00 (paperbound). *Geoarchaeology*, 5(3), 288–290.
64. Pettinger, T., (2021). Capital intensive – Economics Help. [online] Economics Help. Available at: https://www.economicshelp.org/blog/glossary/capital-intensive/ (accessed 9 September 2021).
65. Quigley, N., (1987). Concepts and themes in the regional geography of Canada. By J. Lewis Robinson. *New Zealand Geographer*, [online] *43*(1), 46–47. Available at: https://onlinelibrary.wiley.com/doi/abs/10.1111/j.1745–7939.1987.tb01213.x (accessed 10 September 2021).
66. River Styles. (2021). Backswamp – River Styles. [online] Available at: https://riverstyles.com/?Portfolio=backswamp (accessed 9 September 2021).
67. Rumney, T., (2005). The study of agricultural geography. Lanham, Md.: Scarecrow Press.
68. Sciencedirect.com. (2021). Aggressivity – an overview | ScienceDirect Topics. [online] Available at: https://www.sciencedirect.com/topics/engineering/aggressivity (accessed 9 September 2021).
69. Sciencedirect.com. (2021). Bank Erosion – an overview | ScienceDirect Topics. [online] Available at: https://www.sciencedirect.com/topics/earth-and-planetary-sciences/bank-erosion (accessed 9 September 2021).
70. Sciencedirect.com. (2021). Carbonation – an overview | ScienceDirect Topics. [online] Available at: https://www.sciencedirect.com/topics/agricultural-and-biological-sciences/carbonation (accessed 9 September 2021).
71. Senger, L., & Chang, S., (1975). A matter of terminology—the ecological approach in geography: a re-examination of basic concepts. *Geoforum*, [online] *6*(2), 164–167. Available at: https://www.sciencedirect.com/science/article/abs/pii/001671857590086X (accessed 10 September 2021).
72. Simpson, L., (2002). Geography conversion tables: a framework for conversion of data between geographical units. *International*

Journal of Population Geography, [online] *8*(1), 69–82. Available at: https://onlinelibrary.wiley.com/doi/abs/10.1002/ijpg.235 (accessed 9 September 2021).

73. Taketa, R., (1996). Using Field Sketch Mapping to Teach Basic Mapping Concepts in Elementary School Geography. *Journal of Geography*, [online] *95*(3), 126–130. Available at: https://www.tandfonline.com/doi/abs/10.1080/00221349608978704 (accessed 9 September 2021).
74. Thefreedictionary.com. (1979). Capital-Labor Ratio. [online] Available at: https://encyclopedia2.thefreedictionary.com/Capital-Labor+Ratio (accessed 9 September 2021).
75. Thyne, And Thyne, M., (2008). Concepts in geography through map and photo. *New Zealand Journal of Geography*, [online] *51*(1), 28–28. Available at: https://onlinelibrary.wiley.com/doi/full/10.1111/j.0028–8292.1971.tb00486.x (accessed 9 September 2021).
76. Toyooka, S., Maruyama, R., Toyooka, K., mclerran, D., Feng, Z., Fukuyama, Y., Virmani, A., Zochbauer-Muller, S., Tsukuda, K., Sugio, K., Shimizu, N., Shimizu, K., Lee, H., Chen, C., Fong, K., Gilcrease, M., Roth, J., Minna, J., & Gazdar, A., (2002). Smoke exposure, histologic type and geography-related differences in the methylation profiles of non-small cell lung cancer. *International Journal of Cancer*, [online] *103*(2), 153–160. Available at: https://pubmed.ncbi.nlm.nih.gov/12455028/ (accessed 9 September 2021).
77. Underhill-Sem, Y., (2001). Maternities in? Out-of-the-way? Places: epistemological possibilities for retheorising population geography. *International Journal of Population Geography*, [online] 7(6), 447–460. Available at: https://www.readcube.com/articles/10.1002%2Fijpg.241 (accessed 9 September 2021).
78. Unwin, T., (1997). Book Review: Concepts in human geography. *Progress in Human Geography*, [online] *21*(2), 287–288. Available at: https://journals.sagepub.com/doi/10.1177/030913259702100217?Icid=int.sj-abstract.similar-articles.1 (accessed 9 September 2021).
79. Vocabulary.com. (2021). Cascade – Dictionary Definition. [online] Available at: https://www.vocabulary.com/dictionary/cascade (accessed 9 September 2021).
80. Walford, N., (2001). Reconstructing the Small Area Geography of Mid-Wales for an Analysis of Population Change 1961–95. *International Journal of Population Geography*, [online] 7(5), 311–338. Available

at: https://onlinelibrary.wiley.com/doi/abs/10.1002/ijpg.229 (accessed 9 September 2021).

81. Welch, R., (1978). Conflict, Apparent and Real: A Review of Concepts of Space in Geography. *New Zealand Geographer*, [online] *34*(1), 24–30. Available at: https://onlinelibrary.wiley.com/doi/abs/10.1111/j.1745–7939.1978.tb00874.x (accessed 10 September 2021).
82. Williams, B., (1980). An evaluation of shallow groundwater movement in a small boulder clay catchment in holderness, M. Bonell, University of Hull, Department of Geography, Miscellaneous Series No. 18, (1978). No. of pages: 133. Price: £2.60. *Earth Surface Processes*, [online] 5(2), 197–198. Available at: https://onlinelibrary.wiley.com/doi/10.1002/esp.3760050211 (accessed 9 September 2021).
83. Witherick, M., Ross, S., & Small, J., (2021). Modern Dictionary of Geography. [ebook] Available at: https://cloudflare-ipfs.com/ipfs/bafykbzacedam4qsxqonjzknhkjdl4 gehpxulhqnr2piu4cqvssiwem3pxrhyw?Filename=%28Student%20Reference% 29%20Michael%20Witherick%2C%20Simon%20Ross%2C%20John%20Small%20-%20A%20 Modern%20Dictionary%20of%20Geography-A%20Hodder%20Arnold%20Publication%20%282001%29.pdf (accessed 9 September 2021).
84. Wolf, J., Stanton, M., & Gellott, L., (2010). Critical Thinking in Physical Geography: Linking Concepts of Content and Applicability. *Journal of Geography*, [online] *109*(2), 43–53. Available at: https://www.sciencegate.app/document/10.1080/00221340903582289 (accessed 9 September 2021).
85. www-das.uwyo.edu. (2017). Banner clouds. [online] Available at: http://www-das.uwyo.edu/~geerts/cwx/notes/chap08/banner.html (accessed 9 September 2021).
86. www.biologyonline.com. (2021). Carrying Capacity Definition. [online] Available at: https://www.biologyonline.com/dictionary/carrying-capacity (accessed 9 September 2021).
87. Wynn, G., (2015). Key Concepts in Historical Geography. *Journal of Historical Geography*, [online] *48*, 73–75. Available at: https://www.sciencedirect.com/journal/journal-of-historical-geography/vol/48?Sdc=2 (accessed 9 September 2021).

Index

D

E

I

J

K

L

M

N

O

P

R

S